A Raging Gold Bull or a Hibernating Golden Bear? Investment Reality Check

Over the past two years, all the pundits and so-called "experts" botched their forecasts. While most were predicting that gold prices would reach $2,000 per ounce in late 2012 and early 2013, prices actually fell by 38% from their 2011 peak. In late May 2014, more "experts" emerged from their holes in the woodwork to predict that prices would fall to $840. If you are like most gold investors, you would probably love to kick most of these so-called "experts" in their teeth!

A couple of years ago, billionaire, George Soros, proclaimed that there was a gold bubble and unloaded his massive holdings in GLD, a gold exchange traded fund. Other investors and hedge fund managers dumped their gold holdings and switched their funds into the hot stock market of 2013, where they chased high-tech darlings and junior bio-tech stocks with virtually no earnings but oodles and oodles of promise.

This book concedes that some air may have been let out of the gold bubble but it has not burst, otherwise it would be back to the $350 level again.

The gold exchange traded funds dumped huge amounts of the metal out onto the market in response to sales by disenchanted investors. But somebody bought the gold! Who were they; how much did they buy and why? Those are very interesting questions and this book will attempt to answer them because they may hold the key to where the future lies.

This book talks about some of the key factors that influence the price of gold. It presents a reality check. It looks at the battlefield to try to see where the bulls and bears are located.

Then, the reader will have to decide whether a raging gold bull is about to charge onto the field or whether the golden bear is just hibernating or even worse...dead from effects of an Arctic Vortex.

Everything is not doom and gloom. The fact remains that over the past 10 years, ending in May 2014, the price of gold has risen from about $350 to about $1,300. This is a 271% increase or about 27% per annum. The question is whether the next ten years will show a similar increase or will gold erase those gains and fall back to $350 an ounce?

The title is biassed. A raging bull implies that gold is going to surge upward; whereas a hibernating bear implies that gold is just slumbering and that it will wake up to see another sunny day. So, this book takes the position that gold isn't dead...it's just a question of how quickly it will revive and if there are some bear traps along the way!

Written by a former mining analyst, under a pseudonym, this book covers some of the less publicized factors which influence the price. These will be of interest to gold bugs as well as those who are curious about investing in the yellow metal.

A Raging Gold Bull or a Hibernating Golden Bear?

Investment Reality Check

Published by:

PRODUCTIVE PUBLICATIONS

By

Learn2succeed.com

Incorporated

ISBN: 978-1-55270-703-6

Written by:
Learn2succeed.com Incorporated

Published in Canada by:
Productive Publications, P.O. Box 7200
Station A, Toronto, ON. M5W 1X8
Phone: (416) 483-0634 Fax: (416) 322-7434
Canadian Web Site: *www.ProductivePublications.ca*
American Web Site: *www.ProductivePublications.com*

Front Cover Art:
Spider web adapted from copyright free clip art from *Corel Gallery 200,000*
Corel Corporation, 1600 Carling Ave. Ottawa

Library and Archives Canada Cataloguing in Publication

A raging gold bull or a hibernating golden bear? A reality check / by Learn2succeed.com. Incorporated.

Issued in print and electronic formats.
ISBN 978-1-55270-703-6 (pbk.).--ISBN 978-1-55270-704-3 (pdf).--
ISBN 978-1-55270-705-0 (epub)

1. Gold–Purchasing. 2. Investments. I. Learn2succeed.com Inc., author

HG293.R34 2014 332.63 C2014-904523-9
C2014-904524-7

Disclaimer

This book mentions the names of a number of publicly traded companies, software publishers as well as private businesses. These are provided only by way of example and are not a solicitation to make investments in them; purchase or sell such securities or to purchase the products they offer. Readers should consult a qualified financial advisor before making any investment or purchase decisions.

No representation is made with respect to the accuracy or completeness of the contents of this book and both the author and the publisher specifically disclaim any implied warranties of merchantability or fitness for any particular purpose and in no event shall either be liable for any loss of profit or any other commercial damage; including but not limited to special, incidental, consequential or other damages.

CONTENTS

INTRODUCTION
The Comeuppance of the Talking Heads

Chapter 1
The ETF Bull Went Charging Off in the Wrong Direction

Chapter 2
Where Did All The Gold Go?

Chapter 3
The Hibernating Golden Bear in India

Chapter 4
Gold as a Currency

Chapter 5
Reserve Currencies

Chapter 6
Paper Fiat Currencies

Chapter 7
Where Gold is Traded

Chapter 8
The Government Debt Crisis

Chapter 9
The Influence of Oil on the Price of Gold

Chapter 10
The Demand Side of the Equation

Chapter 11
The Supply Side of the Equation: Gold Mining

Chapter 12
Investing in Gold Mining Companies

Chapter 13
Investing in Gold Shares Listed on a Stock Exchange

Chapter 14
Short Covering Gold Rallies

Chapter 15
Mutual Funds Which Specialize in Gold

Chapter 16
Investing in Gold Coins, Bars and Wafers

Chapter 17
The Gold Futures Market

Chapter 18
Conclusion: The Reality Check!

INTRODUCTION

The Comeuppance of the Talking Heads

Almost All The "Pundits" Have Been Wrong!

It's sickening! Almost all the so-called "pundits" have been wrong in predicting the price of gold. In September 2011,when gold was about $1,921 an ounce, I remember one well respected analyst with Barclays with the best track record who was predicting $2,000 by the end of that year. I seem to recollect many other "pundits" who were predicting that gold would surge to $10,000 an ounce. Well, it didn't happen! Prices fell! Yes, I'll admit that prices did rebound a bit in 2012 but then collapsed to under $1,180 in June 2013.

At that time, more "pundits" with excellent track records emerged from the woodwork and predicted that prices would drop to $1,000 in 2014 and $840 in 2015. As of early June 2014, that downward movement hasn't happened....yet!

Why have so many talking heads on television been wrong? If you have been an investor in gold, wouldn't you just love to kick them all in the teeth?

So, first off, I don't claim to be a pundit. Secondly, I value my teeth and my two gold crowns. Yes, I said "gold" and I paid for them when the price of gold was way up so that certainly wasn't a good investment–at least not yet!

The "Bubble" Didn't Burst Because it Didn't Really Form Properly in the First Place

In 2012, I wrote the book *Gold Investing for Beginners: An Opportunity for Huge Gains or a Bubble About to Burst?* Well, for sure the "Huge Gains" didn't happen but then the "bubble" didn't burst either, however, some of the air was certainly let out. Just for the record, I'll repeat some of my concluding remarks.

Be Sceptical About What People Say!

When your cab driver or barber gives you a tip to buy the same gold stock: watch out!

When you receive a phone call from a stock promoter extolling the virtues of some penny gold stock that will make you millions: watch out!

When you are at a cocktail party and a voluptuous blonde, wearing a skin-tight dress is telling you about a penny stock that will go ballistic, take your eyes off her plunging neckline that goes down to her ankles and reveals a gold ring in her navel. If you're a guy you probably don't have to be told to watch out, because you're not paying attention to anything she's saying! If you are a woman, you're also probably not paying attention because you are thinking "how disgusting" or "I wish I had a bod like that!" or "what a disgusting way to invest in gold!"

When it comes to the stock market; don't pay attention to what people look like or to what they are saying!

Even "Experts" Can Be Horribly Wrong!

Financial analysts who follow companies will make earnings forecasts. Thompson Reuters has a service which lists analysts expectations. This is quite expensive but some online discount brokers (such as Scotia Online) will provide you with a summary at no extra charge of the number of buy, hold and sell recommendations when you look up a stock.

Just because financial analysts are regarded as experts, does not mean that they make perfect predictions. I know of one major Toronto-based brokerage firm that put out a glowing report on Southwest, a gold prospecting company with properties in China. It supposedly had a large deposit but when the drill cores were checked, it was found that the results had been tampered with. Needless to say the shares fell out of the sky and investors lost a ton of money.

Another was Bre-X Minerals of Calgary, Alberta, which supposedly had a large gold deposit in Indonesia that turned out to be a hoax. Unfortunately, most analysts were fooled and, as I recollect, the only one that suspected that something was wrong was fired from his position for refusing to go along with the crowd!

The predictions of many analysts turn out to be incorrect. In this regard, I'll give you a brief quote from Princeton Economist, Burton Maikeil's book *A Random Walk Down Wall Street*: "A blindfolded monkey throwing darts at a newspaper's financial pages could select a portfolio that would do just as well as one carefully selected by experts."

Having said all this, I will grudgingly admit that a few analysts make very good predictions!

Look for an Economist With One Hand!

Former US President, Harry S. Truman, in a fit of frustration, was quoted as saying: "Give me a one-handed economist! All my economists say 'On the one hand, on the other.'" If you listen carefully, you'll find that they haven't changed much over the years!

If you listen to PBS Newshour Economics Commentator, Paul Solman, you will find a healthy dose of scepticism when it comes to economists' ability to predict the future. In fact, many of their predictions are just plain wrong!

I'll tell you a secret. I have an excellent "economist"in the form of my commercial printer. He tells me very honestly when things are bad or good because he prints a lot of advertising material for small businesses in his area. He has a much better feel of the pulse of the real economy than some of the "experts" in skyscraper offices with their heads in the clouds!

Maybe, you can do one better than Harry S. Truman and look for a dead economist who has given up predicting anything!

The Expert's Cop Out--Invest for the Long Term!

One thing that really gets my back up is the so called experts and investment advisors who always creep out of the woodwork after a bad market and make a pronouncement such as "this downturn should not affect the longer term outlook." What they are really saying is that they blew a lot of money....the expert's cop out!

In my view, you have to try to take advantage of a fluctuating market.

Bubbles are Driven by Greed

I don't need to tell you about the dot com bubble which lasted from 1995 to 2000 and involved speculation in technology companies; many of which had no earnings and probably never had any prospect of earning a dime. Needless to say that when the bubble burst, it caused an almighty stock market crash.

In the more recent sub-prime real estate bubble, adjustable rate mortgages offered to millions of Americans (from 2004 to 2007) with poor credit ratings were then bundled ("securitized") and sold with high credit ratings to financial institutions. When the initial low rate mortgages reset to higher rates, many people were unable to afford them; had their homes foreclosed and this also left many financial institutions holding mortgages that had gone sour. In turn, these financial institutions ran into difficulties and, when Lehman Bros. collapsed, it almost brought down the entire banking system in the developed world. As you know, the stock market crashed and the economy went into a tailspin, as credit for business seized up.

There have been many bubbles in the past and if you want to know more about them I would highly recommend that you read a book by Charles Mackay which was first published in 1841. Yes, that's not a typo: it's 1841. It is entitled *Extraordinary Popular Delusions & the Madness of Crowds*. In it, he describes the Dutch Tulip Mania of the early 17th century and excessive speculation in the South Sea Company which traded with South America in the 18th century and which ruined many fortunes when the bubble burst.

More recently, billionaire George Soros, called gold the "ultimate asset bubble" and cut his holdings in the metal by 99%. Many (including myself) disagred. At that time I thought that we may be at the start of a bubble developing but I feel that it's far from bursting.

Bubbles have burst before. It happened to silver after the Hunt brothers attempted to corner the silver market and their scheme unravelled. It happened in 1324 AD when the ruler of the Mali Empire in West Africa was making his way to Mecca on the haij pilgrimage. His camel train was laden with so much gold from his homeland that when he stopped over in Cairo, he spent so much of it that it depressed the price for the next ten years!

To go back to the comments I made earlier, when your cab driver, hair dresser and "in-people" at cocktail parties can only talk about gold–then it's time to watch out! I don't think we are quite there yet. At cocktail parties, I'm still looking out for the shapely lady with the gold ring in her navel! Believe me, I've looked hard....but I haven't spotted her yet!

I ended up by saying: "All I can say is, if a bubble is developing, enjoy the ride and try to get out before it bursts!"

A Raging Gold Bull or a Hibernating Golden Bear?

In this book, I'm going to try to avoid predicting the longer term outlook. With all the pundits and so-called experts having botched their forecasts, I am not prepared to add my name to the list! Instead, what I would like to present is a reality check. Let's look at the battlefield and try to see where the bulls and bears are located. Then, I'll have to leave it up to you to decide whether a raging gold bull is about to charge onto the field or whether the golden bear is just hibernating or even worse...dead from effects of the Arctic Vortex.

All I can tell you is that for the past 10 years ending in May 2014, the price of gold has risen from about $350 to about $1,300. This is a 271% increase or about 27% per annum. Not a bad return! If you change the time frame to the ten years ending March 2012, you would have made a 463% increase or about 46% per annum, while the price of gold rose from $293 to $1,651. On the other hand, you could have bought at the peak around $1,800 and you would now be licking a $500 wound. The bulls and bears are still slugging it out.

Some air may have been let out of the bubble, but I don't think it has burst otherwise it would be back to the $350 level again. So, what I will try and do in this book is present you with, what I think, are some of the key factors to consider as an investor in gold.

I will admit that the title is biassed. A raging bull implies that gold is going to surge upward; whereas a hibernating bear implies that gold is just slumbering and that it will wake up to see another sunny day. So, I don't think gold is dead...it's just a question of how quickly it will revive and if there are some bear traps along the way!

Chapter 1

The ETF Bull Went Charging Off in the Wrong Direction

What is an Exchange Traded Fund (ETF)?

An Exchange Traded Fund (ETF) trades like a stock on an exchange, but it tracks either a commodity, a bond or a share index in the form of a basket of stocks. The first ETFs were created about twelve years ago and since then they have developed into an attractive vehicle for investing. Indeed, in the US alone, there is approximately $1 trillion invested in ETFs. There are now several thousand ETFs and they mostly trade on the NYSE Arca, the Toronto Stock Exchange and the London Stock Exchange.

Some commodity ETFs trade the actual commodity; others trade futures contracts for the commodity and still others a group of stocks that produce that commodity.

ETFs that track commodities (such as gold) are sometimes referred to as ETCs or CETFs where the "C" stands for "Commodity."

Physical Gold Exchange Traded Funds

Probably the best known and most widely traded gold ETF is the SPRD Gold Trust which trades on the NYSE Arca under the symbol GLD. The shares are very liquid i.e., you will always find a buyer or a seller because there can be tens of millions of shares traded in a single day. Indeed, in August 2011, when the gold market was quite hot, over $10 billion worth traded in one day. That's more than the annual gross domestic product of many small nations i.e., the total output of goods and services they produce during the course of one full year!

The shares of GLD are backed by physical gold which is held in safe keeping. Every share represents about one-tenth of an ounce of gold. Gold is added to the fund when there are more buyers than sellers and sold when the reverse is true. At the end of 2011 the fund held 40.3 million ounces with a value of $63.5 billion! By the time I wrote this in Mid-May 2014, those holdings had fallen to 25.1 million ounces with a value of $32.6 billion. This dramatic fall caused the price of gold to tumble since the ETF was forced to sell gold into the market in response to the sell-off in shares. Lower prices caused fear among investors who sold their shares, thereby perpetuating this vicious selling cycle.

Actually, GLD shares hit a low at the end of June 2013 and again in late December. At that time, the talking heads were predicting further declines, and when they did not materialize, the price went up with help from the Ukraine crisis but have since subsided as tensions with Russia "appear" to have abated.

Some Factors to Consider When Purchasing Gold Exchange Traded Funds

One of the biggest advantages of holding a physical gold ETF is that you don't have to take delivery of the gold you buy and you don't have to pay for storage. You can buy and sell in small or large quantities at any time the stock markets are open. A gold ETF lets you play the commodity and removes several levels of risk such as poor management decisions by gold producers, political and tax uncertainty in foreign lands, labour unrest, natural disasters including mine cave-ins, etc.

The disadvantage is that you won't receive any dividends on your investment (although I must admit that gold producing companies tend to pay pretty miserly dividends; if any).

There are a couple of other factors you should consider in owning a physical gold ETF. Firstly, the management fees charged (although modest) mean the fund can sell gold in order to cover the fees and such sales reduce the value of the underlying gold. Theoretically, the longer you hold the fund, the more you are "bled" to death. Secondly, US investors holding positions for over one year are taxed at a higher capital gains rate since the ETF is regarded as a "collectable." For this reason US holders may want to exit prior to a 12-month hold period.

Other ETFs Which Deal In Physical Gold

There is also a physical gold ETF which trades in Canadian dollars on the TSX under the symbol CGL and the price for the bullion, which is quoted in US Dollars is hedged in Canadian Dollars. This is the iShares Gold Bullion Hedged ETF. In this case, it tracks 1/100 th. of an ounce of gold expressed in Canadian dollars at the going exchange rate. I should point out that the trading volume is a fraction of that for GLD, but with daily volumes of between 2-14 million shares, it is still quite liquid.

If you are trading on the London Stock Exchange, the Source Physical Gold P-ETC (symbol SGLD) may be of interest. The physical gold is held in the vaults of the J.P. Morgan Chase Bank in London and the value is quoted based on the London afternoon gold price fixing. At the time of writing, it held over $1.8 billion US dollars worth of bullion.

Index Tracking Gold Exchange Traded Funds

Probably the best example of an Index Tracking Gold Exchange Traded Fund is the ETFS Securities fund which tracks the DJ-AIG Gold Sun-Index and it trades on the London Stock Exchange under the symbol BULL.

Another example is the BMO Junior Gold Index ETF which tracks the Dow Jones North America Select Gold Index and owns 34 mid-cap and junior gold mining companies. It trades under the symbol ZJG on the TSX and the last time I looked at it in detail, it had almost 80% of its gold company holdings located in Canada; with the balance in the US.

It's difficult to make a good comparison between a physical fund such as GLD and an index fund such as ZIG because in the latter, shares of different companies may be added or dropped depending on the index administrators. The second factor is that the shares of many mining companies fail to reflect the true price of gold in a rapidly moving upmarket and yet get disproportionately clobbered during a downturn, such as a drop in the metal price combined with year-end tax loss selling. Thus, from early September 2011 till the end of the year, the ZJG dropped about 29% whereas GLD only dropped about 16%.

Leveraged ETFs

If you want to travel in the fast lane, you can trade leveraged ETFs which give you the promise of two or three times the bang for your buck! The reality is that they are great for short-term trades but not so good for the longer term.

The longer you hold a leveraged ETF for a volatile index, the further it may get away from doubling or tripling the index on which it is based. This is simply because, during a strong bull market the fund manager has to buy at ever increasing prices to achieve his leverage; whereas during a market that is falling sharply, he ends up having to sell into a market that may be reluctant to buy his stock and he sells at distressed prices. In other words, he is getting whipsawed into buying at exaggerated prices during an upswing and at distressed prices during a downturn. This can lead to significant trading losses with the result that a leverged ETF can fail to deliver on its promise.

You can also use leveraged commodity ETFs such as gold, which I will cover next.

Leveraged Gold Exchange Traded Funds

I am going to take you to Las Vegas! You could purchase a leveraged fund such as Pro Shares Ultra ETF (UGL) which trades on the NYSE Arca and attempts to deliver twice the performance (up or down) of gold prices on a daily basis. Actually, it does not hold physical gold but attempts to achieve double the price movements by using futures contracts, options, and swap agreements.

In reality, it lives up fairly well to its reputation according to my brief analysis for the period I looked at from early September 2011 till the end of that year. During that time, the shares dropped about 33% or twice that for an unleveraged ETF e.g., GLD.

Trading in these sorts of vehicles is not for the faint of heart because you can either make a lot of money or lose a lot in a relatively short period of time.

The Enormous Movement Out of Gold ETFs

Billionaire, George Soros, who was one of the founders of GLD, decided it was time to cash in his chips and he sold 99% of his holdings. Hedge fund managers and other holders also decided to abandon ship and to re-allocate their funds to shares in the booming US stock market of 2013. All of this resulted in huge amounts of gold being dumped out into the market. Indeed, a staggering 879.8 metric tonnes (28.3 million ounces) was sold, according to figures compiled by GFMS, Thompson Reuters, The London Gold Fixing and the World Gold Council. In just two days, in April 2013, gold took a huge hit and fell the most since 1980. But...somebody bought it!

Where did all that gold go? That's an interesting question and I'll attempt to answer it in the next chapter.

Chapter 2

Where Did All The Gold Go?

That's an Interesting Question!

Rising incomes in China, combined with gift giving associated with the Chinese New Year and Valentine's Day, pushed China's consumer demand for gold up by 43% to 894.6 metric tonnes for the 12-month period ended March 31, 2014, according to GFMS, Thompson Reuters and the World Gold Council. That figure is a staggering amount and is greater than the amount of gold sold by ETFs during the whole of 2013. It's also significantly higher than the amount of gold mined in China during 2013.

Put another way, Chinese consumers are sucking up all the gold being sold by US investors in addition to all the gold they can mine for themselves.

There's More to the Chinese Story!

According to the International Monetary Fund (IMF) and the World Gold Council, the largest officially reported gold holdings of various countries, as of January 2014, were as follows:

USA	8,133 tons
Germany	3,387
International Monetary Fund (IMF)	2,814

Italy	2,451
France	2,435
China	1,054
Switzerland	1,040
Russia	1,015
Japan	765
Netherlands	612
India	557
Turkey	508
European Central Bank	502
Taiwan	423
Portugal	383
Venezuela	368
Saudi Arabia	323
United Kingdom	310

You will note that two of the Eurozone countries which have serious debt problems are on this top-18 list; namely Italy and Portugal. Also, since I last recorded this list in December 2010, China has more than doubled its holdings and both Russia and Turkey have added significantly more.

So, what's going on?

This more than doubling by China suggests that it is making a serious attempt to place its reserves in gold rather than fiat paper currencies. It is interesting that an article by Ailen Sykora on January 17, 2014 in Kitco News suggested that the Chinese holdings may be much larger than those officially recorded since much additional purchasing has been done by the China Investment Fund, a sovereign wealth fund and, as such, would not appear in the "official holdings" of the People's Bank of China. Indeed, Sykora suggested that total Chinese gold holdings may be as high as 2,710 tons which would rank the country as the world's fourth largest holder.

If Sykora's assertions are correct, this would mean that there has been more than a five-fold increase in China's gold holdings in the short space of approximately three years. Now, that screams out that China does not trust the US Dollar and may be hedging its bets in light of its huge holdings of American debt; now estimated to be about $1.3 trillion in US Government Treasuries; although it masochistically continues to accumulate more. That may be for appearances' sake because China doesn't want to shoot itself in its foot. But realistically, what's a "mere" $1.3 trillion in treasuries compared to its $104.6 trillion in "unofficial" holdings of gold; based on $1,200 an ounce? It's small change!

The Chinese Gold Jewellery Trade

Gold jewellery demand in China has outpaced India's for most of the time since 2003. A September 2011 article by Frank Holmes in *Forbes Magazine*, claimed that China has become the fastest growing gold jewellery market in the world with 6.6 million brides in 2010, who made gold a part of their ritual. This is due to the rise of a Chinese middle class which is contributing significantly to consumer demand. Another factor is rising inflation in China combined with rising incomes, which is causing the Chinese to invest in gold as an asset to protect their wealth.

It is interesting that the communist government, which micro-manages the country and controls so much of what its population can and cannot do, has done nothing to stop the growth in the gold jewellery trade; regardless of whether purchases are made for adornment or investment. In addition, the government has permitted its banks to hold gold on deposit for its citizens and in March 2014, the Shanghai Gold Exchange announced plans to open a second exchange located in the free-trade zone, thereby allowing foreigners to invest. One interesting feature is that purchases will have to be made in yuan rather than US dollars which indicates that the government is attempting to internationalize its currency.

The Chinese Government is Holding Its Cards Close to Its Chest

Not wishing to upset the world currency applecart, the Chinese Government appears to be playing its cards close to it chest and allowing its citizens to accumulate gold. Meanwhile, agencies of the government are quietly accumulating gold in anticipation of the day when the US runs into severe trouble with its bloated debt and when its paper currency becomes....just paper! I'll take a look at that in a later chapter.

Chapter 3

The Hibernating Golden Bear in India

The Importance of Gold in India

Gold plays a very important role in Indian society.

Gold jewellery purchases in India are traditionally associated with religious festivals starting with the Islamic festival of Eid in August and ending with the Hindu festival of Diwali in October. These are followed with the traditional wedding season. Due to high gold prices, demand was somewhat subdued in 2011 and in March 2012, there were protests against the Indian Government's plan to increase duties on gold purchases, with many stores shutting down to express their displeasure.

Increased Tax on Gold Jewellery

In disregard to the protesters, in September 2013, the Indian Government increased its import tax on gold jewellery from 10% to 15% and that was the third increase in 2013. (At the beginning of the year, the import tax was 4%.) These increases were made in an attempt to help narrow the country's current account deficit. Instead of doing that, it will probably result in a flourishing trade for smugglers, since gold is relatively easy to conceal.

Mandatory Set-Aside of Gold for Jewellery Export

On July 22, 2013, the Reserve Bank of India introduced rules which required that shippers of gold set aside 20% of their shipments for re-export. This draconian measure had the opposite effect when banks and traders suspended imports. It will probably cause the loss of many jobs in the jewellery trade and will cause more harm than good.

It harkens back to the days of India's Gold Control Act which came into effect in 1968 and lasted for almost 30 years. That Act prohibited gold imports, exports, the manufacturing and sale of jewellery above 14 carats.

Could the New National Government Loosen Up on Gold?

Hindu nationalist Narendra Modi's MJP party swept to power in the May 2014 elections and he took office as Prime Minister on May 26. His main pledges were to stamp out rampant corruption and to revive the stagnating economy. There are also hopes that his government will loosen up on the import duties and regulations governing gold.

Chapter 4

Gold as a Currency

Way Back When

The term "love trade" is used to describe the jewellery trade. Insofar as gold is concerned, it's been going on for a long, long time. Indeed, gold artifacts that are some 6,000 years old have been found in modern day Bulgaria.

In 3000 BC, jewellery was being worn in southern Iraq by the Sumer people and when the tomb of Egyptian King Djer was located, it was found to contain gold jewellery. He reigned in 2500 BC.

In my student days, I had the privilege of visiting Cairo's Egyptian Museum. I was wonderstruck at the gilded shrines and the solid gold coffin belonging to Tutankhamun. He died in 1328 BC.

Since gold can be beaten down into very fine sheets; so fine that you can see through them, it can be used in windows to protect against the glare of the sun. Although, I am not too sure that was the primary consideration when The Royal Bank of Canada installed such gold plated glazing in its Toronto head office. I suspect that there was more than a bit of ostentatious "Tutankhamunism" about it!

The last Shah of Iran had gold toilet seats installed in his private jet.

Kings and Queens throughout the ages have worn gold crowns as symbols of their majestic power.

I've also got two gold crowns....in my teeth!

The English Cashier Said: "That'll cost a Soverign, love"

Indeed, she may well have said that in the days of Charles Dickens. Back then, the Sovereign was a small gold coin which was valued at one pound sterling. They were minted in the United Kingdom from 1817 until 1917 and again in 1925. Various Commonwealth countries also minted them from 1957.

In 2009, The Royal Mint released a Quarter Sovereign. Obviously, the more recent mints are for collectors or held for investment purposes so you won't see anyone paying for their groceries with one.

Gold as a "Store" of Value

The present use of gold coins is for coin collectors but more importantly for investors. However, there are other ways of investing in gold; for example you could purchase gold bars, wafers etc. You can invest in a physical gold exchange traded fund (as I discussed earlier in this book) which can be bought and sold on a stock market. You could also buy shares in companies that mine the stuff. Alternatively, you could buy gold futures or invest in a mutual fund that specializes in the metal and I'll talk about these other ways of investing later in this book.

I'll also talk about some of the factors which influence the daily price of bullion. After all, the whole purpose of this book is to show you how to invest in gold and hopefully, help you make some money but also alert you to some of the dangers of a possible bubble developing.

When Gold Coins were First Used as Money

According to the National Mining Association, gold was first used as a currency around 1500 BC when the people in Nubia, in modern-day Egypt, began using gold as a medium for international trade.

Around the same time, a coin made of about 2/3 gold and 1/3 silver made its appearance in the Middle East. This was known as the "Shekel."

In 1091 BC, gold in the form of small square coins were legalized as a form of money in China.

The first coins of pure gold were made in the ancient Kingdom of Lydia, located in Asia Minor. That was about 610 BC.

In the year 50 BC the Romans began issuing gold coins called the "Aureus."

Jump many centuries later, and in 1284 AD, Venice (at that time a major maritime trading power) issued the gold Ducat. The English, not to be outdone, issued the Florin; their first gold coin, in the same year.

In 1782 AD the Coinage Act in the United States defined the US Dollar as being a bi-metallic coin which was mostly silver but with slightly over 6%

gold. This was followed in 1900 with the Gold Standard Act which was designed to establish gold as money carrying a fixed exchange rate with other countries.

Two things caused the unravelling of gold as a "standard." They were the debasement of gold coinage and finally the abandonment of the "Gold Standard."

Debasement of Gold Coinage

Pure gold is a very soft metal. If you bite on it, your teeth dent marks will show. In the past, this was used as a quick test to determine if a coin was made of pure gold or not.

In practice, many "gold" coins contain other metals in order to increase their hardness. The most common additive was silver, although copper was also used. However, by decreasing the amounts of gold in a coin, rulers were also able to debase their currency.

Probably the most famous, or infamous, debasement of gold coinage as a currency occurred many years ago. According to Professor Joseph Peden in a lecture delivered in 1984, entitled *Inflation and the Fall of the Roman Empire*, debasement started after the reign of the Roman Emperor Augustus, when gold coinage was circulated with a weight of 45 coins to a pound of gold. Under Emperor Caracalla, the gold content was reduced to 50 coins to a pound and further reduced to 60 during the reign of Diocietian, who ruled from 284 AD to 305 AD. Diocietian also debased the silver coinage.

In a desperate effort to bring inflation under control, Diocietian also introduced wage and price controls, however, his efforts failed and by the time he abdicated, inflation stood at close to 100%.

Modern Debasement of Coinage

In a twist of irony, the US Mint in 2012 was facing a similar Diocietian problem, where inflation in the price of metals has caused one cent coins and the nickel coins to have higher metal content values than they are worth. As a consequence, the Obama Administration is seeking the permission of Congress to debase the currency! Meanwhile, Canada has buried its penny!

The Introduction of Banknotes

Commercial banks in England began issuing banknotes which were paper money and could be redeemed at the issuing bank in return for gold or silver. Similar banknotes were issued by banks in the United States and the fad caught on, so much that at one time there were as many as 5,000 different types of notes in circulation.

Unfortunately, many banks issued more notes than gold and silver held to back them. A sudden lack of confidence in the paper money could cause customers to rush to their bank in an attempt to convert their paper money into the precious metals. If a bank had insufficient gold or silver to hand out to their customers, they were forced into "bankruptcy."

In 1694, the Bank of England was granted the sole right to issue banknotes in England. The United States followed more than 200 years later, when in 1913, the Federal Reserve Bank was granted similar rights. Theoretically, the paper currencies that were issued could be converted into gold or silver. In practice, people did not ask for their paper to be converted and based their trust in the credibility of their respective governments.

The Gold Standards

In 1900, the United States passed the Gold Standard Act which committed the United States to maintain fixed exchange rates with other countries. This was followed in 1913, when Federal Reserve Notes were required to be backed by 40% in gold. It was suspended during World War I and partially reintroduced in 1934, when President Roosevelt reduced the value of the dollar, by setting the price of gold at $35 per ounce, but convertibility was restricted to foreign banks. To make sure that American citizens could not demand to convert their paper money into gold, in the previous year, he prohibited them from owning gold coins, bars or certificates of gold ownership. A further draconian measure was introduced in 1942, when Roosevelt issued a presidential edict to cause all American gold mines to be closed. In 1961, the American Government forbad Americans from owning gold in foreign countries, as well as domestically.

The British Gold Standard Act was passed in 1925 and this permitted the conversion of banknotes into gold bars but not into gold coins. The Act was suspended in 1931 because of large outflows of gold. Essentially, the "Standard" was found to be unwieldy and unstable.

In 1931, major commercial banks, in both Austria and Germany failed after a run on the gold that backed their currencies.

Bretton Woods

Bretton Woods was a conference which took place in July 1944 at which the world's major industrial powers agreed to establish a set of rules for commercial and financial relations between them. The rules were not fully implemented until late in 1958.

The Bretton Woods Conference was an attempt to find a way to rebuild the world's economic system after World War II. It led to the creation of the International Monetary Fund (IMF), together with an organization which was the predecessor to today's World Bank.

Another important decision at Bretton Woods was to require the participating countries to tie the exchange rates of their currencies to the US dollar. It also permitted the IMF to provide temporary bridge financing in the event that there was an imbalance of payments.

Bretton Woods collapsed in 1971, when President Nixon abruptly terminated the ability of foreign governments to ask for their holdings of US dollars to be converted into gold. This was the beginning of modern fiat paper money, which I will talk about more in Chapter 6.

Chapter 5

Reserve Currencies

Floating Exchange Rates

After the collapse of Bretton Woods, many countries allowed their currencies to float freely in value, while others remained pegged to the US dollar.

The Concept of a Reserve Currency

A "Reserve Currency" is generally regarded as one that is stable and can be used to price goods and commodities. Obviously, the US dollar is today's major reserve currency since, gold, silver, oil and most other major commodities are quoted in the greenback. In other words, if you are in Japan or Europe you would have to convert your Yen or Euro into US dollars in order to buy gold on an American Exchange. It is also used by banks to pay off international debts or to influence the foreign exchange rates of their currencies.

With the horrendous debt of the United States, there is much talk about creating alternative reserve currencies. For example, several major oil producing states in the Middle East would like to create a reserve currency based on the price of oil. Russia and China would like to disentangle themselves from having the greenback as a reserve currency.

A Futile Call for the Return of the Gold Standard

A number of people, including former US President Regan, have called for a return to the Gold Standard. They argued that it would help prevent politicians from the willy nilly printing of more and more mountains of paper money and would thereby restore some discipline to the budgetary process.

The challenge would be to establish a price for gold. If it is set too high, the US Treasury could be faced with truck loads of gold being dumped on its doorstep asking to be converted into paper money. If the price was set too low, the Treasury could quickly be depleted of its stash of gold bars.

There's another problem. As pointed out in a September 14, 2011 article by Alix Steel in *The Street*, in order to cover US dollars in circulation in the form of bills, savings, deposits etc., (the "M2 money supply") gold would have to sell at $35,000 an ounce. This, in spite of the fact that the US Government is already the largest single holder of gold in the world with 8,133 tons of the stuff stashed away at Fort Knox! In spite of this, it would also have to purchase additional gold every second of the day to fund its voracious debt.

As you can see, the system is already out of control and any return to a gold standard is impractical at this late stage of the game.

The alternative, would be to establish a two-tier currency; one based on the current greenback and another with paper notes that can be converted into gold.

Ironically, free market forces have already created a pseudo gold-based currency in the form of physical gold Exchange Traded Funds (ETFs) which I covered earlier. These trade like shares on a stock market but their price tracks the price of gold and they are backed by gold holdings held in a vault. I have to point out that even though you cannot use a share in an ETF to purchase your groceries, it represents a small baby step in the direction of creating a currency based on the price of gold.

A Two-Tier Gold Price

A few years prior to the collapse of the Bretton Woods accord, a two-tier pricing system emerged. One was set at $35.00 per troy ounce for transactions between central banks and the other was a free floating rate based on market supply and demand. Actually, the demand was so great that in 1968, the London Gold Market had to be closed for two weeks because of the ensuing rush into gold.

As I mentioned previously, three years later President Nixon prohibited the United States from converting foreign owned US dollars into gold, causing the collapse of the Bretton Woods accord.

The Devaluation of the US Dollar

Accompanying the American Government's refusal to convert its paper currency into gold, the paper dollar was devalued in 1971 by repricing its value in gold from $35.00 per ounce to $38.00 per ounce. A further devaluation took place in 1973 when the "official price of gold" was set at $42.22 per ounce.

The US dollar came under intense selling pressure and later in 1973, major currencies were allowed to float freely. Meanwhile, the free market price in London had shot up to more than $120 per ounce.

The Legalisation of Private Individual Gold Holdings

At the end of 1974, American citizens were allowed to own gold in forms other than jewellery. In 2004, the Chinese Government permitted its citizens to buy gold and further encouraged them in 2011, by making it easier for its citizens to buy and hold gold with their banks.

Call for a Co-operative World Monetary System

A couple of years ago, Robert Zoellick, then head of the World Bank, made a call for a "co-operative world monetary system" which would basically involve the US Dollar, the Euro, the Yen and gold as a global reserve currency ("Breton Woods II").

Chapter 6

Paper Fiat Currencies

What is a Fiat Currency?

No! It's got nothing to do with automobiles!

Actually, Wikipedia has a great definition. The term is derived from the Latin "fiat" which means "let it be done" or "it shall be [money]." Essentially, money is created by government decree.

The first fiat money was used in 11th century China. Much more recently, in 1971, after President Nixon disallowed the convertibility of the United States dollar into gold, the US dollar together with all other world currencies have basically become "fiat currencies".

In plain speak, the government takes a piece of paper, prints the head of a president, prime minister, monarch, whatever, on it and decrees that this piece of paper is worth so many dollars; pounds sterling, euros etc. You can no longer go to the government and demand a certain amount of gold in return for your paper currency.

What Will Replace the World's Fiat Currencies Over the Course of the Next Century?

The United States, as the world's largest economy is in a quagmire of debt. Japan as the world's third largest economy is also struggling under a massive government debt burden. There are fears that China, as the world's second largest economy, is hiding the huge internal debt problems of its provinces, municipalities and state run industries under a corrupt communist rug.

The counties on the peripheral of Europe still face horrendous challenges. These are unkindly referred to as the "PIIGS" i.e., Portugal, Italy, Ireland, Greece and Spain.

It's reaching the point where all the taxes raised by some countries will go towards paying interest on their debt; leaving nothing for social security, healthcare, armed forces etc., and leaving nothing to pay down the principal on the debt. In order to provide these basic government services, the countries will be forced to borrow more. It's a spiralling debt crisis which will be very hard to escape from.

Confidence is the "glue" that holds this shaky, debt ridden economic world of ours together but that confidence is waning as seen by the growing acceptance of gold as a way central bankers are using it to hedge their bets. But gold is not the perfect answer in a digital world and a digital economy. It's cumbersome to transport in a fast-moving e-commerce society where transactions take place over the Internet in a matter of seconds.

What is a Cryptocurrency?

A cryptocurrency is a peer-to-peer digital currency; the creation of which is based on cryptography, which makes it difficult to counterfeit. It also permits any transactions to be validated, so that some crypto magician can't just pull them out of a hat.

"Yea! So what's cryptography?" Well, in simplistic terms, it is a method of secret writing which uses code or ciphers. It owes its origin to the Greek "kryptos" meaning "hidden." What's interesting is that over 3,000 years ago, the ancient Egyptians were using hieroglyphics to encrypt secret messages. In more recent times, Hitler used it to communicate with his top generals and it was only the genius of a small team at England's Bletchley Park that were able to decrypt the messages; thereby shortening the war by several years, although their efforts went largely unrecognized and unrewarded.

There are various kinds of cryptocurrencies but the leader is undoubtedly Bitcoin. Its units are known as "bitcoins" (note the capitalized "B" for the concept and the lower case "b" for the units.) Instead of using a bank, you can store your bitcoins in your digital wallet on your desktop, laptop or smartphone. You can even get an Android app which let you incorporate a Bitcoin Wallet in your smartphone.

Bitcoins are developing into a global virtual currency and provide an easy, quick and inexpensive way to make worldwide money transfers without incurring any absurdly excessive bank transfer fees. In addition, they enable users to avoid being gouged by their banks on their money exchange rates, where they help themselves to a portion of the funds. Of course, this treads heavily on the toes of the almighty commercial banksters and threatens to overturn their tables as money changers. As a consequence, they have cold-shouldered bitcoins and would dearly love to have central banksters do their dirty work and crucify the perpetrators.

Of course, this puts government servants and central banksters in a quandary. Suddenly, they are plunged into the midst of a populist uprising which worships a new world cryptocurrency and they find themselves impotent to do what regulators are supposed to do....that is: regulate! Does this mean that the world's slaves to paper fiat currencies can throw off their shackles and embrace a democratic digital currency which can fearlessly transcend international borders and let them purchase anything in the Bitcoin universe or simply keep their bitcoins stashed in their digital wallets, in a hope that they grow in value?

Not so fast! Enter former Fed. Chairman Alan Greenspan who famously warned of "irrational exuberance" in the stock market and correctly predicted the dot-com bubble. Unfortunately, he didn't have the same insight into the lead up to the US sub-prime mortgage bubble which was growing at an alarming rate during his watch. In his latest epiphany, the retired Chairman has declared that bitcoins are a "bubble." Is he right?

There's certainly a Bitcoin Mania but is it a "bubble?" If you want to read more about it, read *Bitcoin Mania: The Birth of Worldwide Virtual Currency or the Start of an Insane Bubble?* which can be purchased securely online at ***www.ProductivePublications.ca*** in Canada or at ***www.ProductivePublications.com*** in the US or overseas.

Could a World Virtual Cryptocurrency Based on Gold Be Emerging?

Of course, rather than using physical gold, a virtual cryptocurrency **based on** gold presents itself as an interesting possibility. Such a cryptocurrency could be converted into gold and vice versa. It's happened in the past to paper currencies. All you have to do is roll the clock back less than 100 years to a world where paper currencies were convertible into gold on demand.

Fast-forward to today, and there are reports that the British Royal Mint is involved in a plan to mint a limited edition of physical gold coins based on Bitcoin. (I assume with the encryption enclosed on a microchip.) What is even more intriguing is that the jurisdiction involved would be a tiny 3-mile long island called Aldernay, which is located in the English Channel off the coast of France. What is even more interesting is that the island belongs to neither the UK nor to France but is a feudal property owned by the Duke, who just happens to be Her Royal Highness, Queen Elizabeth!

A gold-backed cryptocurrency overcomes one of the main objections to returning to a gold standard and that is that there is insufficient gold around. This could be overcome with cryptocurrency units (like bitcoins) which can be subdivided into tiny units. If those "tiny" units get too high in value, they can be subdivided again, and again and again but you can always convert them back into gold even if it is only for a few tiny grains.

The Problem with a Cryptocurrency Based on Gold

There are some fundamental problems with using a cryptocurrency, like Bitcoin, based on gold.

The first, is that if a cryptocurrency is based on gold, the gold has to be stored somewhere. As such, it is exposed to seizure by government regulators, tinpot dictators or any other hostile party. This is in contrast to bitcoins which can exist secretly and anonymously outside of any jurisdictional boundaries.

The second, is that there are two values involved. There is the price of gold itself, which constantly fluctuates relative to a fiat currency such as the US dollar. There is also the price of bitcoins which fluctuate on their own. Of course, if the value of the bitcoins drops below the value of the gold, an owner could simply melt down the coin and sell it for its value in gold. Conversely, if the value of the encrypted currency greatly exceeds the value of gold, the price of gold would likely come under pressure as gold traders working outside of the realm of cryptocurrencies dump their gold holdings. So, in either event the price of gold would come under pressure and could collapse.

It is hard to conceive of a currency which has two values and for that reason, I question the survivability of any cryptocurrency based on a physical commodity such as gold, silver, platinum, oil, etc. On the other hand, physical currencies based on metals could provide an interesting alternative to fiat paper currencies. Some commodities would not be suitable. For example, I can't see myself walking into my local convenience store and pouring oil from my "oil wallet" into a measuring cup in order to buy a bag of chips and a can of pop.

Could a New World Digital Cryptocurrency Emerge that is Not Based on a Commodity Such as Gold?

Even if a new cryptocurrency is not backed by physical gold or another commodity, it's tantalizing to think that it might emerge as a virtual world currency during the course of the next century. Bitcoin has certainly made a strong start in this direction but whether it can survive the upcoming regulatory maelstrom is another question.

In addition, Bitcoin would have to achieve widespread acceptance among the peoples of all nations. That's a tall order as anyone who has attended a session in the United Nations will be able to confirm!

Maybe another world cryptocurrency will emerge which could be beatified by the governments of the major economic powers. Unfortunately, this may only happen after the fiat currencies of the world collapse under mountains of national debt and that could be a very, very messy affair...so fasten your digital seatbelts!

Chapter 7

Where Gold is Traded

The Main Exchanges Which Trade in Gold

The US dollar, the British Pound sterling and the Euro are the three main currencies used to price an ounce of gold and there are several exchanges where the metal is traded.

The price in London is fixed twice each business day. It is quoted in US dollars, Pound Sterling and Euros. It is a somewhat antiquated system, dating back to 1919, when bullion traders and refiners got together at the St. Swinton's Lane offices of N. M. Rothschild and Sons. This arrangement with Rothschild lasted until 2004, when Barclays Capital took over the task of fixing the gold price and for the two daily fixings. Also, the fixings are now conducted electronically.

At the time of writing, a scandal was brewing over this "cosy fixing" by a small number of financial institutions and there have been allegations that some of them were using inside information to make gold trades during the time while the afternoon fixings were taking place and thereby taking an unfair advantage of the system in order to make money at the expense of other traders.

In 2006, the two exchanges which traded gold in New York City (the Commodity Exchange or COMEX and the New York Mercantile Exchange or NYMEX) merged, however there are two divisions. Gold is only traded on the COMEX division and prices are quoted in US dollars.

The COMEX also introduced futures trading in gold. Under this arrangement, a buyer can purchase gold for future delivery at a specified time and at a specified price. Typically, the contract is sold before physical delivery of the metal takes place.

NY Globex is a 24-hour electronic trading platform for trading gold and other commodities, together with futures contracts and derivatives. It is a world trading system with no geographical borders.

Gold is also traded on several other world exchanges, as follows:

- Shanghai Gold Exchange (SGE)
- Hong Kong Mercantile Exchange (HKMEx)
- Mid-America Mercantile Exchange (Mid-Am)
- Dubai Gold and Commodities Exchange (DGCX)
- Tokyo Commodities Exchange (TOCOM)
- Sydney Precious Metals Exchange
- Australian Bullion Exchange (ABX)
- Singapore Mercantile Exchange (SMX)
- Pakistan Mercantile Exchange (PMEX)
- National Multi-Commodity Exchange of India (NMCE)
- Indian Commodity Exchange (ICEX)
- Nepal Derivatives Exchange (NDEX)
- Commodities & Metal Exchange Nepal (COMEN)

I found it interesting that Dubai now trades about 25% of the world's physical gold. Also, as I mentioned earlier, in March 2014, the Shanghai Gold Exchange announced plans to open a second exchange located in the free-trade zone, thereby allowing foreigners to invest; however, gold has to purchased in yuan. Standard Charter Plc, in an interview with Bloomberg, points out that this will expand the range of investment options for yuan deposits around the world which it estimates at the equivalent of $240 billion USD. This makes me wonder if this could be a preliminary baby-step to getting the yuan accepted as a world reserve currency.

The 24-Hour Trading Market

Gold is traded 24-hours a day, 5-days a week on the major exchanges. A fascinating way to follow it (at no cost) is to visit the Web site: *www.kitco.com*. It tracks the price in London and then as the day progresses, there's an overlap when the New York NYMEX opens. Later in the day NY Globex takes over, then Sydney and finally Hong Kong.

Kitco provides a graph with the current day's price together with the two previous days. The graph lines are in different colours to make it easier to follow. It also provides historical charts plus news releases, press reports and commentary. It's a great way to follow the gold market as well as the other precious metals. It's free!

The Price of Gold in Different Currencies

Because of floating exchange rates, the price of gold can vary quite considerably over time periods of several months to several years. A great way to make comparisons is to go to *www.kitco.com/market*. Here you will find the price of gold quoted in some of the world's major currencies.

What you have to watch out for is the price of gold when there is a major swing in currency rates. I'll take a couple of simplistic examples.

At the time of writing, the Canadian dollar was worth about 91 cents in US dollars. This means that an ounce of gold that costs an American $1,255 in USD would cost a Canadian $1,379 in Canadian dollars (CAD).

It was even worse a number of years ago. When the Canadian dollar tumbled to 76 cents US in March 2009, it would have cost an American about $940 to purchase an ounce of gold, whereas a Canadian would have had to pay about $1,236 in Canadian dollars for that same ounce of gold. The disparity was about 32%.

Put another way, a Canadian purchasing gold in March 2009 and then selling the same gold in March 2012 would have suffered a loss had the price of gold remained constant. Fortunately, for such an investor, the price of gold had risen to around $1,658; an appreciation of about 34% which would have offset the 32% loss and leave a miserly 2% profit.

On the other hand, an American purchasing an ounce of gold in March 2009 in US dollars and selling it three years later, in March 2012, would have made a handsome 34% profit.

What I am trying to convey here, is that in a world of fluctuating currency exchange rates, you have to be very cautious when investing in gold, because exchange rates can either work in your favour or against you.

The Price of Gold Fluctuates with the Strength or Weakness in the US Dollar

When the Eurozone financial crisis started in 2010, there were times when the Euro seemed ready to fall apart. At those times, investors rushed into what was perceived to be a safe haven i.e., the US dollar, and to a lesser extent gold. When the Euro looked as though it might pull through, the Euro strengthened; the US dollar weakened and gold prices fell.

Since those times, when the US dollar is strong, there is a tendency for gold prices to be weak and vice versa. It seems as if everyone was running from one side of the boat to the other in an attempt to keep it upright in a storm.

Gold prices are definitely affected by strength in the US dollar but I wonder aloud how long this will last, especially when the enormous US deficit is taken into account. The US is in a slightly better position than Greece...but not by that much! That will be covered in the next chapter when I discuss government debt to GDP ratios.

Chapter 8

The Government Debt Crisis

Government Debt-to-GDP Ratios

Gross Domestic Product (GDP) is one of the primary indicators used in determining the economic health of a country. It is the total value of goods and services produced over a specific period of time. A crude comparison would be too equate it to the total value of what you have earned during the course of the year from your labour and from your investments.

To calculate the Debt-to-GDP ratio, take the national (or federal) debt of the country and divide it by its GDP. This will tell you how many years it will take to pay off that debt. Again, going back to the personal example, how many years would it take to pay off your personal debts if you used all your earnings?

The Debt-to-GDP ratio will tell you how healthy a country's economy is and it will allow you to make comparisons with the economic health of other countries.

Figures taken from Eurostat (the European statistical agency) with European Community and the CIA's *World Face Book 2010* for the rest of the world, reveal that for the 182 countries that were surveyed, the average Debt-to-GDP for the entire world was 59.3%. It was somewhat higher, at 79.25% according to figures generated by the International Monetary Fund (IMF).

Some of the lowest figures came from the oil-rich states, such as Saudi Arabia and Oman, with under 10%, to countries like Japan with a huge Debt-to-GDP ratio of 220%. The United States came in with a figure of 94.36% for 2010 and an estimated 102% for 2011. The UK came in with figures between 75.5% and 79.9%. Canada had figures in the range of 83.95% to 84.0%.

The ability of a country to be able to repay its debt is critical. If you take a look at the crisis in the European Union; specifically relating to the Eurozone currency, you will see that Germany has a Debt-to-GDP ratio of about 83% and that compares to Greece at about 143%, Italy at 119%, Portugal at 92% and Spain at 60%.

While the US is not the worst, it still has a staggering Debt-to-GDP ratio. This is very worrying since it is the world's single largest economy outside of the European Union. By way of contrast, countries in the developing world, such as China, India, Brazil, have Debt-to-GDP ratios of about half or less than that of the United States.

One thing is certain: these debts have to be repaid unless countries go into bankruptcy. This happened to Germany in the Weimar Republic before Hitler came to power; the Russian debt default of 1998; the Argentine debt default in 2002 and another possible default of Greece in the coming year or so.

These are events which shake the financial world and can cause turmoil in stock markets and bullion markets. Suffice to say, at this point, is that not all government debt and bonds are secure and a great deal depends upon the country that is issuing them. Unfortunately, the safest haven that is perceived, at the time of writing this book is the US, however, the yields are very modest and well below the rate of real inflation. So, you could have safety, but not be able to earn any significant amount of money! Tricky, isn't it!

Debt at the State and Provincial Government Levels

If US, UK and Canadian debt at the national or federal level is a concern, it gets much worse when you dig deeper down to the individual states in the US or the provinces in Canada. Many of the US states are teetering on the verge of bankruptcy. The debt levels in three of Canada's largest provinces (Ontario, Québec and British Columbia) are increasing rapidly and, once again, there appears to be little political will to cut spending or raise taxes. Contrast this with the US, where balanced budget provisions have forced many of the states to make drastic cuts to government spending and to raise taxes wherever possible.

The fact that many states and provinces are almost in as bad shape as Spain or Italy should be cause for alarm and will almost certainly affect future borrowing costs in the debt and bond markets. Increasingly higher yields will not necessarily make up for the risks involved.

The Eurozone Debt Crisis

The Eurozone debt crisis roiled financial markets during 2011 and threatened to continue to do so during 2012 and beyond. Before I dive into a discussion of the crisis, it may be helpful to set the stage.

The European Union (EU) is basically a free-trade and commerce zone which comprises 27 countries; only 17 of which have the Euro as a common currency. The most significant country that stayed out of the Euro is the UK. Also, the Scandinavian countries and Hungary remain outside.

Of the 17 countries that use the Euro as a common currency, Germany has by far the largest economy and accounts for about half of the Eurozone's GDP. France, Italy and Spain are the next largest. As I mentioned earlier, a big challenge has been the PIIGS (Portugal, Ireland, Italy, Greece and Spain)., Some of these smaller countries which are in trouble, such as Greece, Ireland and Portugal, between them only account for about seven or eight percent of the total Eurozone GDP. Against that, Italy and Spain account for a fair chunk and what is even more worrying is that France appears to be slowing down and growth in Germany is also cooling.

The crisis started with Greek government debt that got completely out of hand because of high wages paid to civil servants, generous pensions to government workers and taxes that were not being collected. As expenditures outstripped income, the country borrowed and borrowed; while at the same time hiding the truth from its Eurozone partners. Finally, the cat jumped out of the bag as the country struggled to avoid bankruptcy.

The trouble is the contagion effect. French banks had bought a lot of Greek government debt, so if Greece goes down and even with the restructuring by issuing new bonds. If the bonds become worthless, some major French banks would be exposed to huge potential losses which pushes us back to images of the Lehman Bros. failure which rocked world markets after the sub-prime mortgage bubble burst.

In spite of all the efforts that have been made to save Greece, it is still not clear whether the country can avoid eventual bankruptcy. The severe austerity measures have pushed the economy into a tailspin and this reduced the tax base since people are losing their jobs, companies are closing down and the whole economy is turning into one gigantic mess.

One very painful solution is that if Greece goes bankrupt and leaves (or is forced out of) the Eurozone and returns to using the Drachma as its currency, you can expect stock markets not to take kindly to such news. On the other hand, it may cause a flight to gold by Europeans who will be seeking safety from a currency that will likely be in decline.

Both Ireland and Portugal have bitten the bullet and appear to have their financial houses in order. The same is also true of Spain where the government has taken strict measures to correct the situation; even though the country faces a staggering 23% rate of unemployment. Having said this, if Greece goes under, investors will be looking for the next shoe to fall and all of these countries could be at the mercy of negative market sentiment and punitively high interest rates when they have to refinance their upcoming debt repayments.

Italy was a major worry for the markets during 2011, since it is a major power in the Euorzone and is basically deemed too big to fail. Former President Berlusconi was long on words and short on taking concrete action. Berlusconi, now performing social work as a result of being found guilty of fraud, was later replaced by a bureaucrat with significant smarts when it came to European finance. He was later replaced by Matteo Renzi in the elections which followed.

One way to determine how well this theatre is playing out is to examine the borrowing costs for these countries. It seems that 7% is the tipping point. There's no hard and fast rule but it is generally regarded that rates much above this figure place an onerous burden on countries to repay their debt in the future.

All I know for sure is that the Eurozone crisis is still with us and is likely to have a major influence on stock market behaviour and gold prices during 2014 and beyond.

A Recession in Europe?

Leading up to May 2014, there were signs that the entire European economy was slowing and European banks were still reluctant to loan money to businesses. In fact there was danger of deflation where prices drop. People cut back on their spending because they figured that they can purchase goods cheaper if they wait for prices to fall. Deflation is the opposite of inflation, where people rush out to buy goods before the prices go up.

In early June 2014, European Central Bank (ECB) chief, Mario Draghi, cut interest rates to record lows in an attempt to stop deflation. He even imposed negative interest rates on overnight deposits in an effort to force commercial banks to lend money businesses.

Draghi has also hinted at starting a bond buying program to target some of the weaker economies and provide some form of quantitive easing, but he is likely to run into stiff resistance from Germans who feel that they should not be bailing out the economies of countries that have been mismanaged.

Could a Recession in Europe Cause a Hard Landing in China?

With inflation, especially in food prices, running high in China and India, both countries have slammed on the financial brakes in an attempt to slow growth. Unfortunately for them, they both rely on exports to fuel their continued growth since their internal populations are not yet at a stage where they can consume all that is produced.

With a possible slowdown in Europe, which is China's largest export market, together with a lackluster recovery in the US, then China's economy could be headed towards a hard landing and if that happens there could be significant social unrest which could make the Tiananmen Square massacre look like a picnic party. Obviously, Communist Party leaders are very anxious to keep their jobs, their heads together with all their perks.

At the time of writing, in mid-May 2014, the Chinese real estate market was in a slump with new construction down by as much as 22%. This is quite worrisome since new construction had been a major contributor to economic growth in the past. In turn, this has led the central government to call on major lenders to speed up the signing of new home mortgages.

Another factor is higher wages in China. The World Bank found that wages in China have risen by 19% since 2000; compared to 4% in the US. This has profound implications, since many American and European manufacturers moved their operations to China in order to take advantage of cheap labour, but that cost advantage is rapidly disappearing. Higher fuel prices have caused increased freight costs to transport finished goods from China. When both factors are added together, the pressure to repatriate operations to the

US homeland has increased. Also, all the fracking activity has let to a US domestic glut of cheap natural gas and oil which is a boon to energy-intensive manufacturing operations such as steel production.

Having said that, China's economic growth rates are nothing to be sneezed at. At the time of writing, the economy was growing at about 7.4% which is a far cry for the 17% of a few years back but still very respectable compared to the anaemic growth in America and Europe.

The problem is that civil unrest is a major concern in China, as workers demand better working conditions, even higher wages and benefits. Indeed, in 2010, according to the latest figures released in China, there were 180,000 protests; mostly relating to working conditions and wages. If this is the "official" figure, what is the unofficial figure?

In addition, many peasant farmers have been thrown off their land to make way for development and have not been properly compensated. Corruption among the communist elite is rampant with the result that many Chinese now harbour resentment against the wealthy. All of this is a tinderbox for future revolts.

What does all this mean for gold? So long as the Chinese middle class and wealthy continue to make money, they will spend it on gold jewellery to flaunt their wealth and gold bars as an investment and as a hedge against geopolitical events spinning out of control.

If China falls on its economic face, then gold could certainly come under pressure as holders are forced to sell in order to provide the basic necessities such as food and fuel.

US Housing Prices and the Inventory of Mortgage Foreclosed Homes

In May 2014, there was still an inventory of one American million homes which were under water i.e., the amounts owed on mortgages were greater than the value of the underlying real estate.

Higher mortgage rates and stricter lending standards are also placing downward pressure on home sales. It is difficult to generalize across the entire country, but home prices have more or less stabilised after an initial rebound, but they are still well below pre-bubble prices. Consequently, the wealth effect, when people feel that they have tons of money to spend has been dampened. Since 70% of the US economy is made up of consumer spending it has serious implications for the economy as a whole.

You may well ask: what has this got to do with the bullion market? Actually, it is positive, since the price of gold benefits from bad news on the housing market! The economy slows and the Fed comes to the rescue and prints more money.

Here we go again...or maybe not!

The Unemployment Factor

At the time of writing, US unemployment was running at 6.3% in May 2014. This is the official figure but many argue that the rate is much higher at around 14%, when you add the number of people that have given up looking for work and the huge number of people who are underemployed i.e., with high qualifications but are working at a low paying job (such as flipping hamburgers in a hamburger joint).

On the other hand, new auto sales appear to be very strong but this could be due to the fact that the average age of a vehicle is about 11 years and after a while, old Betsy has to be replaced before it rusts into oblivion. It is hard to predict how long this pent up demand will continue to influence the market. Unfortunately, manufacturing in that area is highly automated with the extensive use of robots so it may not move the needle much on new employment rates.

Even if the unemployed are receiving some kind of government support, this usually runs out after a while. Also, those without work tend to be very frugal in their spending habits which puts a damper on consumer spending and, in turn, has a negative effect on the economy.

The Multiplier Effect

When a high paying industrial job disappears due to a plant closure, there are frequently other jobs that are dependent on that industrial worker. There could be a store keeper who sells his goods to the worker, contractors who renovate houses may no longer get work, the services of gardeners and lawn maintenance people may no longer be required.

This is an inexact science, but frequently the loss of one high paying industrial job could result in three more jobs being lost in a community.

If people lose their jobs and can't pay the mortgage, they may lose their house and if it can't be sold, it may remain vacant. The municipality will loose the tax base and have to lay off workers to remain solvent.

It's a nasty cycle!

Consumer Debt

According to the Federal Reserve, average American consumer debt increased in the first quarter of 2014 and stood at 5.75% and amounted to $3.1 trillion. This is well up from the $2.6 trillion in 2009, in the aftermath of the financial crisis. Obviously, the average American consumer is living well beyond his or her means and paying interest rates on debt that significantly reduces their real worth.

The situation in Canada seems to be a lot worse, according to figures released by Statistics Canada, in the second quarter of 2014, which indicated that average Canadian households are carrying debt of 163% of their disposable income. This leaves Canadians very exposed to any future economic downturns.

According to the BBC, for the UK, consumer debt reached a record high of £1.43 trillion in November 2013. This figure includes mortgages and if you strip them out, the ratio of debt to annual income has actually declined from 167% at the start of the financial crisis to 140%.

Whichever way you cut the cake, it is obvious that the interest being paid on personal debts and mortgages is seriously eating away at real wealth. The reality of the situation is that high interest rates on debt are grinding consumer's noses into the dirt and consumers have to do something drastic on a personal level if they are just going to keep their heads above water.

Helicopter Ben and "Stay-the-Course" Yellen

The collapse of the US housing bubble and the bankruptcy of Lehman Bros. nearly caused a meltdown of the entire world financial system. The US economy was heading towards the ditch and the Federal Reserve had cut borrowing rates to the bone. Fed Chairman, Ben Bernanke, had a couple of other rabbits to pull out of his hat. The first was to flood the market with money! Indeed he joked that, if needed, he would fly around in a helicopter and throw bags of money out of the window.

This became known as "QE I" or the first phase of Quantitative Easing. The stock markets loved it and showed their appreciation by climbing in value. The gold market loved it. Essentially, the money printing press was being cranked up, which in turn debased the value of the US dollar.

When QE I ran out of steam, Bernanke pulled another rabbit out of his hat: QE II. Here, he set about buying 10-year treasury bonds to help keep yields down and replace them with short-term treasury bills. Again, the gold market loved it.

Towards the end of 2012, the sun had set on QE II and Helicopter Ben pulled QE III out of his hat which included the purchase of mortgage-backed securities. Once again, the gold market loved it.

One thing is clear. In spite of Bernanke throwing his figurative bags of money out of his helicopter, it is apparent that most of it has become stuck in the trees. The banks have benefited from it but very little has floated down to the general public at ground level. In spite of fed funds rates of between zero percent and 1/4 percent, have you found your "friendly banker" offering you loans carrying interest at 2% or your "friendly credit card company" reducing the interest on your balance from 22% to 2%? I will admit that mortgage rates have come down but are much more difficult to get because of tightened lending standards. Basically, the Federal reserve is saving the hides of its commercial banking friends with little regard for the "small guy"and that's why QE I, QE II and QE III have not been particularly effective at the consumer level. On the other hand, the gold market loves it when the US currency is debased in this manner!

Just before Helicopter Ben left office he started cutting back on the $85 billion per month in bond and mortgage purchases. He reduced the purchases by $10 billion at his last Fed meeting and his successor, Janet Yellen has continued to reduce at the same rate at each of her Fed meetings since she took office on February 3, 2014.

The question is: what happens when QE III comes to an end? Will the Fed raise interest rates? At the time of writing in mid-May 2014, it seemed as if a soft job market and relatively tame core rate inflation would likely keep the Fed from raising rates until early 2015. One could argue that all these actions by the Fed have simply prevented the economy from collapsing and one may well ask, what will happen after this $1 trillion per year level of support has been completely removed? Will the economy be strong enough to stand on its own two feet or will it go into reverse again?

One thing is obvious. Gold prices thrive on bad news. So, if the recovery falters, gold investors will probably do quite well. If the economy manages to continue its anaemic expansion or spurts forth, it will likely put downward pressure on gold prices.

Higher interest rates tend to cause money to be withdrawn from the gold market since physical gold holdings do not pay interest. Also gold costs money to store in a vault. On the other hand, it is easy to become myopic and overly concentrate on the Fed and the US market when the main market drivers are China and India.

US Political Gridlock

The second half of President Obama's first term in office has been characterized by legislative gridlock between the Republican led House of Representatives and the Democratically controlled Senate. I don't want to jump headfirst into the angry hornet's nest of American politics, but, as an outsider from Canada, it appears that a disconnect has developed between the US political elite, hemmed in by high paid lobbyists, representing major business groups, and the constituents that they are supposed to represent. Many of their constituents are unemployed and have lost their homes with a result that poverty rates are rising at an alarming rate. According to the latest census figures available, the US Census Bureau stated that 16% of the population in 2012 lived in poverty.

The divide between rich and poor has widened considerably. This resulted in the "Occupy Wall Street Movement" which appeared to represent a general disgust of the political establishment, together with what was perceived as corporate greed. The movement may have died down but resentment still runs deep and at the time of writing there were minimum wage riots when McDonald's held its annual meeting on May 22, 2014 which led to 138 arrests.

There appears to be no clear leadership of these large groups of disgruntled people and it wasn't clear to me how they can topple a well entrenched political establishment with extensive police and military power at their disposal. On the other hand, the Arab Spring started without any clear leaders! This is a harrowing thought for stock market investors but could also be a positive for gold, since it could become a safe haven for those wishing to flee the US dollar!

Horror of All Horrors: a Repeat of 9/11

The horror of all horrors are repeats of 9/11, the train bombings in Madrid; the subway bombings in London or the Bali nightclub bombing.

Let's take 9/11 as an example. The American stock exchanges were closed for several days and when they reopened, the price of shares plunged and it took almost another year for them to recover.

Threats are everywhere. Some attempts have been thwarted such as the shoe bomber's attempt to blow up a transatlantic flight, the Yemini Taliban's attempt to blow up two passenger airliners with explosives hidden in printer toner cartridges being transported as air freight.

Unfortunately, many threats remain. Nuclear terrorism is high on the list with the relative ease of detonating a "dirty" bomb that could spread radioactive material over a major city's downtown area, causing havoc for many years thereafter. Another terrifying threat is the 12,000 or more shoulder-mounted anti-aircraft missiles that went missing from arms dumps as the Gadaffi regime fell in Lybia. Many of these are quite likely to now be in the hands of al quaeda operatives.

Osama bin Laden may be dead and many of his followers killed but al quaeda is not finished. It is like a beast with many heads, and for every one you chop off, it grows ten more. For sure there will be more violence and if reports are true, economic disruption seems to be high on the new menu of horrors. The timing is very difficult to predict but as an investor, you have to hope for the best but be prepared for the worst.

Obviously, the US and Israel are the top prime targets for further terrorist attacks. So, anything which undermines the US could have negative effects on its currency and could lead to higher gold prices. It is not pleasant to gain from the misfortune of others, but let's face reality and put the cards on the table. During 9/11, gold prices in London shot up 32% from $215.50 per ounce to $285.00.

The Core Rate of Inflation-an Economist's Absurdity

The core rate of inflation is something that has been concocted by economists. The rate takes the real rate of inflation i.e., the rate of increase in the cost of goods and services that we consume and strips out the key components of energy and food. I personally find that this is an absurdity and I've yet to find an economist who has been able to survive without eating, driving his car or heating his home.

Unfortunately, the core rate of inflation is something that central bankers dote over in making their magic predictions and in setting interest rate policies. Again, I find this is an absurdity.

In April 2014, the core rate of inflation in the US was 1.8%. In Canada, the latest figure was for March 2014 and it stood at about 1.3%. In the UK the core rate was somewhat higher at 1.7% in March 2014.

The rate of inflation will steadily eat away at your wealth unless you're able to offset it with higher returns, which will also have to make up for the taxes you pay.

The Real Rate of Inflation

If you add back the factors that are stripped out in calculating the core rate of inflation, you will end up with a much higher figure. In February 2012, Statistics Canada released figures which showed a 6.5% increase in the price of energy (oil, gas, electricity, etc.) over the 12-month period ending January 2012. Consumers also paid 4.9% more for food purchased in retail stores on a year-over-year basis.

If you add back the factors that are stripped out in calculating the of inflation, you will end up with a much higher figure. Indeed, a feature on CNBC *Fast Money* claimed that the real rate of inflation in the United States was closer to 10%. Another estimate by *Shadowstats* claimed that the consumer rate of inflation in the US is around 6%. Let's split the difference and say it's 8%.

Since the price of oil and food commodities are global in nature, I think it's reasonable to assume that similar rates of inflation occur elsewhere in the world, insofar as "non-core" rate items are concerned.

The Inflation Adjusted Price of Gold

I have to admit that I was always under the impression that inflation was good for the price of gold, however, when I was researching the topic for this book, I found, much to my surprise, that gold generally lags inflation. Indeed, according to a June 2010 article by Michael Kosares published by *GoldSeek*, the inflation adjusted price of gold would have to be over $7,500 an ounce!

Having said this, there are extreme circumstances when gold became a very viable hedge against inflation. Kosares points to the hyperinflation in Germany during the 1930s during the Weimar Republic. It led to the collapse of the Reichsmark and anyone who had converted their Reichsmarks into gold before the hyperinflation took hold, would have been very, very handsomely rewarded had they wished to convert back into the paper currency. However, since the currency ended up as worthless, the best they could say is that they preserved their wealth by buying gold.

This collapse of the currency still haunts many Germans today and is probably the reason why they are so reluctant to let the European Central

Bank (ECB) turn on the paper currency printing presses; unlike the Americans!

The other side of the coin is that the Eurozone countries may be forced into expanding the money supply if they are to save the indebted nations of Southern Europe and Ireland. Such a scenario would be positive for gold prices although, as I said earlier, there appears to be a lot of German resistance to the idea.

Chapter 9

The Influence of Oil on the Price of Gold

Oil Prices Should Be Viewed as a "Tax" on the Industrial World

Since oil is used so widely in our society to heat homes, produce gasoline to fuel automobiles, trucks delivering goods, farmers using tractors and other machinery to plant, tender, collect and deliver their crops, its cost has a major impact on our economic well-being. In addition, petroleum products are used to fly the aeroplanes that cross our skies, the bunker oil to fuel the ships that cross our oceans and the "stuff" that fuels many of our electrical power plants. I am sure that you can think of many other uses.

Basically, almost everything in our modern consumption-oriented society is dependent upon oil and prices of the goods and services we consume are dependent on its cost.

Indeed, many of you may remember the oil embargo of 1973-1974 when the Organization of Arab Exporting Countries (OAPEC) consisting of the Arab nations of OPEC plus Egypt, Syria and Tunisia, embargoed the export of oil to the United States in retaliation for the decision by the US to re-supply the Israeli military during the Yom Kippur War. It only ended after the US was successful in getting Israel to withdraw from the Sinai Peninsula and return the land to Egypt and also to withdraw Israeli troops from some parts of the Golan Heights in order to address Syrian concerns.

I am not here to pontificate on who was right and who was wrong: you can decide that for yourself, but US consumers were faced with line-ups at the gas pumps and the US and world stock markets went into a severe down turn.

In 2008, oil prices reached an all-time high of $145 per barrel and the economic house of cards collapsed. The stock market crashed and it wiped out trillions of dollars of wealth in the matter of a few months. On the other hand, gold prices hit a new high of around $1,000 in March 2008 (up from about $640 a year earlier); only to drop back to below $800 later in 2008.

Based on this, if we have another oil crisis, it is likely that gold prices will also increase substantially.

The Arab Spring

The 2011 Arab Spring took the world by surprise. It started with the suicide of a street vendor in Tunisia who had been denied a permit to sell his produce. It toppled a brutal regime, and the ruler's wife was reported to have fled the country with 1.5 tonnes of ill-gotten gold. This was followed by uprisings in Egypt, Libya, Yemen, Bahrain and Syria. Only Tunisia seems to have any apparent success in toppling a brutal dictator. At the time of writing, the situation in Egypt seems to be under a military-civilian "democratic" control, Yemen remains unstable and Syria is engaged in a civil war.

Syria is not a major oil producer and the same holds true for Yemen, however, unrest in the region makes investors very nervous because of any contagion to neighbouring oil producers; especially the world's largest, Saudi Arabia.

Like it or not, it's all about oil; that black gold that fuels our economies and without it we would be plunged into an economic ice age in which our economies and stock markets would freeze to death! On the other hand, gold could be a beneficiary!

Oil from the Middle East and North Africa (MENA)

Since Europeans are more dependent on Libyan oil than the US, prices on UK Brent Crude market went up to $118 per barrel when the Kadafi regime came under attack and oil supplies were significantly curtailed.

The economic well-being of the United States and the UK are highly dependent upon the price of oil. Since Saudi Arabia is the largest producer, many governments have to grovel, appease and bow to Saudi interests.

So, you definitely need to pay attention to what is affecting oil production around the world and what could derail any economic recovery in the 2014-2015 period.

The Arab Spring uprisings have placed fear among the royal elite in Saudi Arabia of the possibility of similar happenings in their Kingdom. As a result social programs such as building schools, housing and hospitals have been instituted to appease the downtrodden masses. These programs have a cost and some people have estimated that Saudi oil will have to sell for around $90 per barrel just to pay for them.

The (slightly) good news is that it is not in the Saudi's interest to see the industrialized world go into a steep recession and cause the demand for oil to fall. Therefore, they are unlikely to let prices rise too much; assuming of course that they can control the situation, because their capacity to pump huge amounts of more oil onto the world market in response to a crisis is somewhat limited.

Could Iran Still An Cause Oil Crisis?

In late March 2011, stringent sanctions were placed on Iran in an attempt to get it to cease its nuclear bomb program. A preliminary agreement led to some loosening of those sanctions but the tough work of coming to a permanent agreement still remains, in spite of the fact that there are a few encouraging signs of de-escalation.

The Europeans have basically stopped buying Iranian oil and insurance has been denied to oil tankers which carry Iranian oil. In spite of the sanctions, both China and Russia have continued to support the Iranian regime.

In the past, when the rhetoric got hot, Iran threatened to block the Straits of Hormuz, the 17 kilometre wide passage at the exit from the Persian Gulf through which 17% of the world's oil supplies pass, including a large portion of Saudi Arabia's production. In the past, the American Navy has dismissed these threats as well as Iranian naval exercises as futile propaganda; however, the reality is, that if fanatical elements of the elite Iranian Revolutionary Guard employ suicide tactics, it could cause disruption and plunge the world into another major oil crisis, with devastating effects on economic growth accompanied by a hard hit to the major stock markets of the world. On the other hand, gold prices would probably rise.

Another factor is the attitude of the Israelis towards any nuclear bomb development by Iran. Any pre-emptive military strike by a very hawkish Israeli leadership will surely plunge the US and other countries into a serious Middle East War which could have very serious consequences on oil prices and, in turn, the economic well-being of most of the industrialized world. Share prices would be negatively affected in a big way and gold prices could be positively affected.

The Sectarian Divide in Iraq and its Possible Split-Up

Another potential oil problem is Iraq.

Personally, I am not sure if the invasion of Iraq achieved any significant security for future supplies of oil from that country, in spite of all the blood, matériel and dollars squandered. As American troops made their final withdrawal, at the end of 2011, the sectarian divide in Iran deepened with Shiite President Nuri al-Malaki's attempt to arrest the Sunni Vice-president, Saleh al-Mutiak, on charges of terrorism after he described al-Malaki as a dictator who is "worse than Saddam Hussein."

At the time of writing, there was certainly no brotherly love on display and there were accusations that Shiite Iran was behind attempts to curtail the Sunni power base in Iraq. This inflammatory rhetoric threatens to plunge the entire country into a sectarian civil war which could drag in neighbours such as Sunni Saudi Arabia, Shiite Iran and the Kurds who want autonomy in the north in spite of Turkey's opposition, since its own Kurdish population would also want autonomy.

Since Iraq is once again a significant world oil producer, this could put supplies in jeopardy and cause world stock markets to fall and gold to rise in price.

Could the Turmoil in Ukraine Cause a Natural Gas Crisis in Europe?

Tsar Vladimir, in his bid to carve himself a place in Russian history books, has demonstrated an acute ability to tell bare-faced lies in denying to German Chancellor Merkel that he had any intention of invading Crimea, two days before he did just exactly that. This is reminiscent of Hitler's antics with the naivety of British Prime Minister, Neville Chamberlain, in his pre-World War 2 meeting.

OK...I had to say that! Back to natural gas! Following the Fukushima nuclear disaster in Japan, the Germans, under Merkel, decided upon, in my view, a very short-sighted and ill-informed decision to discontinue using their nuclear energy electricity generation. Now, Chancellor Merkel has got her knickers (schlüpfer) in a knot! On the one hand, she has cut her nuclear umbilical cord in return for a dependence on Tsar Vladimir for about one-third of Germany's energy needs in the form of natural gas. Now she is placed in the "unenviable position" of bowing and scraping to Tsar Vladimir not to cut off her gas while at the same time being pressured to turn it off in retaliation for his "bad behaviour."

One thing is almost for sure. Ukraine is going to get royally screwed by Tsar Vladimir. Through his chums at Gazprom, Ukrainians are being faced with a huge increase in their natural gas prices; payment of back-invoices and pre-payment for new supplies. Tsar Vladimir is turning on the screws; learned when he was head of the KGB.

What does this mean for gold? Prices ran up sharply during the Crimea crisis and then settled back after the actual annexation. Does this mean that Tsar Vladimir is going to play dead? I doubt it. His ego is going to destabilize Ukraine until he can annex the eastern portion of the country and then he can pick off portions of Romania, Latvia, Lithuanian, etc., which have ethnic Russian populations in "need of protection."

OK...Tsar Vladimir is not a nice guy, so what does this mean for gold prices? Unfortunately, it's positive! In early 2014, gold prices increased significantly during the unrest; only to fall back by about $55 per ounce after it "appeared" that tensions were cooling; even though the situation on the ground suggested otherwise.

Oil From Canada

You may have noticed that I did not mention Canada in the above discussion. The situation there is different since it is a major oil producer with reserves in the Alberta/Saskatchewan oilsands close to the total reserves of Saudi Arabia, the world's largest supplier. Unfortunately, the costs to extract this petroleum from the oil sands is upwards of $40 per barrel as opposed to the $2 to $3 per barrel in the deserts of Saudi Arabia.

Also, Canadian oils sands producers are faced with the challenge of being land-locked. American political games around granting permission to allow the construction of the Keystone-XL pipeline to carry Alberta crude to Gulf Coast refineries have certainly riled Canadians. The problem is further exacerbated by protest from environmentalist and indigenous peoples in British Columbia, through which any pipelines to the Pacific coast would have to pass. As a consequence, more and more oil is being transported by rail which is somewhat more expensive but it does allow producers to get the best price; depending on the destination.

Oil is a Major Factor in Economic Growth

Also, the higher the price of oil; the more attractive alternative sources of energy become such as ethanol, natural gas, geothermal, hydro, wind, solar, etc. Therefore, there is an incentive to try and keep a lid on prices so that the "competition" does not get a chance to "run amuck."

To summarize! The industrialized world, i.e., the economies of western countries plus China, India and the Far East are highly dependent on oil prices. Any increase in prices is like an increase in world taxes–it sucks money out of the economies of the industrialized world.

When US oil prices hit $113 per barrel in April 2011, many economists began to predict a major economic slowdown. I remember one economist who estimated that for the first time since 2008, Americans, on average, were spending more than 10% of their disposable income on fuel and gasoline. In response to the elevated price, the US Government released oil from the strategic oil reserve, which temporally put a damper on prices, although they later went back up to over $100 per barrel. At the time of writing the price of West Texas Crude was selling at around $104 a barrel.

I have discussed this topic at some length in order to demonstrate that a major factor such as this could affect the economy and thereby gold prices. Stay alert to developments!

Gold is Primarily a Crisis Hedge

In conclusion, I'd say that gold is primarily a **crisis hedge** as has been clearly demonstrated in Ukraine. To some degree, it is also dependent on oil prices since those can dictate economic growth and stability or the lack thereof in a crisis. On the other hand, gold only acts significantly as an **inflation hedge** in the extreme circumstances of hyperinflation.

Chapter 10

The Demand Side of the Equation

The Total Amount of Gold Demand

The manufacture of gold jewellery consumes vast quantities of gold. Indeed, according to figures prepared by GFMS, Thompson Reuters and the World Gold Council, about 58% of gold demand, amounting to 2,361 tonnes in 2013, came from the jewellery trade. So, trends in the jewellery trade play an important role on the demand side of the equation and China and India dominate the jewellery market; with China in the lead.

Technology accounted for 408 tons or 10.0% and investments have fallen significantly over the past few years and now only account for about 22% of demand; well down from about 33% only a few years earlier.

I'll examine each of these areas in turn.

The Love Trade

As the Gold Council remarked in connection with the gold jewellery trade: "demand is driven by affordability and desirability by consumers." I'll add, that "desirability" is, in essence, the desire to show off wealth! It is also used in India to pay for dowries.

Actually, when you get down to the nitty-gritties, you will find that jewellery demand in China has outpaced India's for most of the time since 2003. A September 2011 article by Frank Holmes in *Forbes Magazine*, claimed that China has become the fastest growing gold jewellery market in the world with 6.6 million brides in 2010, who made gold a part of their ritual. This is due to the rise of the Chinese middle class which is contributing significantly to consumer demand. Another factor is rising inflation in China combined with rising incomes, which is causing the Chinese to invest in gold as an asset to protect their wealth.

By way of contrast, North American demand for gold jewellery has been soft and was also significantly subdued in the UK and many other markets due to adverse economic conditions.

So, is gold jewellery an investment or an adornment?

In my opinion, it's part of both. When push comes to shove, gold jewellery can be melted down and the gold sold....as a commodity! In the meanwhile, it can make a pretty good investment while helping to adorn people's bods.

Gold's Technology, Dentistry and Industrial Uses

If you exclude jewellery, gold's industrial role is relatively modest. It's used in dentistry, electronics and for making the gold thread called "jari" which is used in fashionable clothing in India.

The use of gold in electronics declined modestly in the 2012-2013 period but still accounted for a respectable 408.6 tonnes in 2013. According to GFMS,

Thompson Reuters and the World Gold Council, in the first Quarter of 2014, the industrial use of gold accounted for less than 10% of gold consumption.

My teeth can attest first-hand to the use of gold in dentistry. I sport two fine gold crowns. The downside is that I have stopped smiling at crooks in self-defence! Having said this, there has been a recent decline in the use of gold in dentistry as newfangled amalgams, cobalt-chrome alloys, ceramics and dental implants have shoved gold off it shiny perch.

Central Bank Purchases Underscore a Fundamental Rethinking of the World Monetary System

Central banks to the rescue!

In the last decade, there has certainly been a U-turn in the attitude of central bankers around the world insofar as the role of gold is concerned; especially as a "pseudo" reserve currency.

The British Government, in a stroke of misguided genius, sold half of its gold reserves between July 1999 and March 2001 at the bottom of the gold market. The average price fetched was $275 per ounce. Compare that with the early June 2014 price of about $1,250 and you will see the astounding folly of the British politicos of the day!

The International Monetary Fund (IMF) decided in September 2009 to sell one-eights of its gold holdings. These were sold to central banks around the world so as to provide funding for low income countries.

Central banks continue to purchase gold and accounted for about 10% of the demand in 2013, according to GFMS, Thompson Reuters and the World Gold Council. This was down just slightly from the previous year, however, it still indicates that central banks continue to be wary of the US dollar as the world's reserve currency. It is obvious that central banks and governments are recognizing gold as a currency; even though it has not been anointed as an "official reserve currency."

Gold's "Currency" Advantage

You can take a gold bar and go practically anywhere in the world and exchange it into local currency (even though you might get some suspicious looks). It's not practical to do the same with a barrel of oil or a truckload of copper. So, gold is a perfect medium for exchange. Theoretically, you could get paid in gold, then purchase goods, services, pay the mortgage, etc. in gold...almost anywhere in the world. Let's face it, gold is acting as an international currency even if it's not officially a "reserve" currency!

Chapter 11

The Supply Side of the Equation: Gold Mining

Gold Alchemists

Even though alchemists throughout the ages have tried to create gold from lead and other metals, all attempts have failed....until relatively recently! In May 2011, there was a report that the Xia lab group at Washington University, St. Louis had managed to transform minute silver nanocubes into gold nanocubes by exploiting their electrochemical difference when in solution. They are a little too small to threaten the US Government's gold holdings in Fort Knox.

Gold Mining

Gold comes from the ground and the total amount that has been mined is estimated at over 174,000 tons. In 2013, gold mining accounted for 71% of the total supply of gold, according to figures prepared by GFMS, Thompson Reuters and the World Gold Council. The world's largest gold producer is now China which surpassed South Africa about seven years ago.

Recycled gold accounts for much of the remainder on the supply side.

Panning, Sluicing and Dredging

Gold occurs in minute quantities in seawater; but you are certainly not going to get rich by trying to extract it.

Since ancient times gold has been mined from the earth. A popular method is to pan for it, by swirling river or other alluvial sands with water in a shallow pan. The gold flakes separate from the sand and gravel. This method was popularized in the Californian Gold Rush of 1848-1855 and again in the Yukon Gold Rush in the Klondike in 1896. Indeed, to this very day you can take gold panning vacations in both of these areas.

A slightly different method was used along the shores of the Black Sea around 1200 BC. Here, river sands were sluiced through sheepskins to dislodge the gold and according to the National Mining Association, this may have been the inspiration for the Golden Fleece of Greek mythology.

Sluicing has become a little more sophisticated since ancient times. Sluice boxes are man-made boxes with "riffles" set in the bottom to catch the gold.

Dredging is a technique where sand or gravel is sucked up from a river or lake bed by a floating dredge above. Gold is then trapped as the material is passed through a sluice box.

Underground Mining

Underground mining really took off after the discovery of gold in the Witswatersrand, South Africa, prior to the Boer War. Some of the deepest underground mines in the world are still operated there today. Thus, the TauTona Mine goes down 2.4 miles (3.9 kilometres) from the surface.

Open Pit Mining

Open pit mining dates back many years. It is said that the Kolar gold fields in India were mined as open pits about 1000 BC. In more modern times, huge pits are excavated using explosives and the material is withdrawn in large trucks and sent for processing.

In many cases, the gold occurs as a byproduct when mining for other metals; especially copper.

Extracting Gold as a Byproduct of Mining for Other Metals

A few companies have specialized in extracting gold from the leftover tailings from mines that are extracting other metals, such as copper. An example was Gold Wheaton which merged with Franco-Nevada in December 2010. In essence, they have the expertise to tease low grade amounts of gold from left-overs and which are not the main focus of many mining operations. They can enter into long-term contracts which are often to their benefit if the price of gold increases.

Extracting the Gold

Gold is a very stable metal and is very resistant to corrosion, however, it does dissolve in cyanide and this is still used today as one of the main methods of separating the gold from its host rock.

Purifying the Gold

Once gold has been cast into ingots at a mine, it still has to be purified. There are two locations which have gained world dominance. The first is in Switzerland near the Italian border and the second is in Dubai.

Storing the Gold

Obviously investors are not going to store bars of gold under their kitchen sinks. So, a whole storage industry has developed with vaults in London, Singapore, Shanghai and Dubai.

Scrap Metal Recycling of Gold

A secondary supply of gold comes from the scrap market. I'm sure you have seen advertisements from businesses which want to buy your old gold jewellery or ornaments. When prices were high, it seemed that every Tom, Dick and Harry was getting into the gold scrap business.

New Gold Supply

At the time of writing, several new mines were coming into production; notably the Detour Mine in Ontario, the Malartic Mine in Québec and possibly Barrick's massive but much maligned and problem-stricken Pascua-Lama mine in Argentina which is running way behind schedule and way over-budget. Several other smaller operations will add to the supply side and offset declines from mines whose deposits have been depleted.

Exploration Expenditures Slashed

The 28% drop in the price of bullion during 2013 has crimped miners' exploration plans as companies slashed expenses. In a January 17, 2014 article in Bloomberg, it was estimated that exploration expenses plunged 30% or by $10 billion in 2013. Also, the share prices of many junior mining companies have been decimated as a result of being unable to raise equity in order to continue their exploration programs. Since it can take a dozen years to get a mine into production. after a deposit has been discovered, this means that future mine supply will be limited and may be insufficient to replace mines whose ore bodies are being depleted.

Chapter 12

Investing in Gold Mining Companies

The Double Jeopardy of Investing in Gold Mining Companies

When investing in gold mining companies, you are exposed to a double jeopardy.

Firstly, you are exposed to fluctuations in the price of gold as a commodity.

Secondly, you are exposed to all the risks associated with getting the stuff out of the ground, local politics, bad management practices, unforeseen disasters such as when the mine manager slips on a banana peel and falls down the mine shaft. I'm speaking figuratively, but you get the idea!

Mostly It's About Grades and Mine Life

When you buy shares in a mining company, your investment is largely based on how many ounces of gold there are in the ground and how much it's going to cost to get them out; and how long the mine is going to last before the gold is exhausted.

Watch Out for Higher Mining Costs!

I talked about the real rates of inflation as opposed to the so-called "core rates" earlier in this book. Unfortunately, the mining industry tends to be at the very high end of the real rates of inflation. Also, skilled workers are increasingly hard to find.

Wages for miners tend to be quite high in developed countries like Canada, the United States, Australia, but quite low in some West African, Indonesian, South American and other countries. So, there is market place pressure to increase the wages of workers in developing countries. "Equal pay for work of equal value" is the premise used in the United States and Canada when dealing with gender wage discrimination in the workplace, but it is also rearing its ugly head in international labour discrimination. Why should a worker in the West African country of Bakino Faso be paid a fraction of the wage for similar work in Canada. Obviously, the wages paid to a worker in Bakino Faso may already be far higher than local wages for work of comparable effort, so the wages might be considered "fair."

I don't want to get into a philosophical argument on this topic, but merely to point out that rising mining costs are a real concern for the mining industry.

Watch Out for Higher Capital Costs!

The costs to open a new mine are inflating at an incredible pace. I don't have any hard and fast figures, but it's not uncommon for a new mining operation to come in well above budget and cost a great deal more than anticipated. Also, when the red ribbon is cut and the "opening switch" is flicked on for the first-time, many unforseen problems may arise. These initial operating glitches have to be ironed out and it may take several months for a new mine to reach its optimum capacity. Meanwhile, impatient investors may be dumping their stock in frustration.

Inherent Risks when Investing in Mining Companies

You only have to read your newspapers to see how dangerous the mining business is; especially when it comes to underground operations. Chilean miners trapped underground and then miraculously rescued; American and Chinese coal miners killed underground. The picture is not pretty.

In Yellowknife, in Canada's Northwest Territories, there was a very nasty incident involving the deaths of miners at the time of a bitter labour dispute at a local gold mine. It happened several years after I had made an unsuccessful attempt to visit the mine, but had to abandon it because of a flight delay caused by a terrorism prank.

In a non-gold mining incident, Cameco's uranium mine at Cigar Lake in Saskatchewan suffered a major underground flood.

What I am saying is that investing in gold mining companies is not without unforseen risks. So, don't put all your eggs in one basket, but diversify your gold portfolio between the shares of a number of companies.

Political and Taxation Risks

The political risk can be very real. Hugo Chavez, the former Venezuelan leader, unilaterally expropriated all the foreign owned gold mining assets in his country. Others were tempted to follow in his footsteps, such as President Morales, the Bolivian leader, however, he has been frustrated in his efforts because the miners in some of his silver mines wanted to keep their well paying jobs.

Of course, "nationalization" can be much more subtle. The left-leaning President Ollanta Humala of Peru has given up his previous calls for outright nationalization of the mining industry. He appears to have recognized the contribution it can make to the economy, however, it has not been a sweet swan song as can be seen when in 2011, gold and copper mining giant, Newmont, cancelled a new billion dollar mining project. I am not privy to all the issues, but there has to be give and take for projects to go ahead and if there is too much "take" and not enough "give" they will stall to the detriment of development in the host country.

An operation does not have to be located in a developing country for problems to arise. Take the Alberta oil patch as an example. Former Premier Stalmech, suddenly decided that he would increase royalties on oil and gas companies operating in the Province. This caused an immediate uproar with share prices falling. It took many months for the Premier to finally see his folly and partially reverse some of his actions. In the meanwhile, many major players cancelled their development plans and pursued interests elsewhere, in neighbouring Saskatchewan, British Columbia and the United States. Because of the long time horizons of many of these projects, by the time the Premier woke up, much irreversible damage had already been done.

The Risks in Growth by Acquisition

Many large gold mining companies seem incapable of discovering new deposits on their own and instead, rely on growth through acquisition of junior companies. Sometimes, they may be tempted to bid at relatively high prices and the results on the stock market may not be pretty.

A classic case is Kinross, which made a couple of takeovers at prices which the street considered excessive to achieve growth. The share price of Kinross was punished for what many perceive as poor management decisions.

Chapter 13

Investing in Gold Shares Listed on a Stock Exchange

Stock Market Investing

Before I start showing you some ways in which mining shares are evaluated, I'll give you a brief tour of stock market investing.

Full-Service Broker

If you set up an account with a full service broker with a major firm, you should obtain access to research reports on the companies their analysts follow, including the economic outlook from their economists.

You should also expect to receive advice on what to purchase and what to sell. These are trained professionals, however, even though they may be trying their best, their advice is not always good and you have to be prepared for recommendations which go wrong.

A full service broker will do all the transactions for you, however, the commissions can be quite high, such as $200 per trade and the amount will increase, the larger the size of the trade.

Discount Broker

A discount broker will perform the trades for you but most will not provide advice or research for you. Essentially, you phone in your order and they do the transaction for you; telling you where the bid and ask prices are and the number of shares available at those prices. They then place the order--end of story.

Because their role is limited to doing transactions, their commissions are substantially less than those of a full-service broker. They will vary from firm to firm but could be about $45 per trade.

Online Discount Brokerage Account

An online brokerage account lets you do all your own transactions from your computer over the Internet. They are set up securely with passwords to prevent unauthorized trading.

The big, big advantage of online brokerage accounts are the low commissions; with many charging less than $10.00 per trade. Most will also offer a live trader for assistance in placing an order, but expect the fees to jump to the regular discount broker fees of possibly as much as $45.00.

Some online brokerage accounts are very basic. They show you the bid and ask prices for the stock you are interested in and the number of shares available for purchase or sale. Once you have placed your order, you can take a look to see if it has been filled or not and at what price. You should also be able to see your account history and obtain your statements online.

Some online brokers offer much more than plain vanilla trading. Thus, Scotia iTrade lets you see in-depth detailed quotes for the next five bid and ask prices. This can be very helpful since you can see where larger quantities of shares are offered for purchase or sale. I regularly use this feature.

Scotia iTrade also offers some basic technical analysis tools as well as access to news releases on the company you are interested in. It will display the major market indices as well as the most active stocks. Other online brokers such as TD Waterhouse will also offer some of these features but charge a higher commission.

The Cash Account

A cash account is a plain vanilla brokerage account. You transfer money to it and use those funds to trade stocks. Most brokerage firms will let you set up automatic transfer arrangements with your bank account.

Foreign Currency Account

Since Canadians frequently want to trade in the US market and to a lesser degree, Americans in the Canadian market, you may want to set up a foreign currency account. Thus, most Canadian investors will have a Canadian dollar account in which they trade stocks listed on the Canadian exchanges and a US dollar account for shares traded on the NYSE Euronext, NASDAQ etc.

One thing you should watch out for is the foreign exchange fees when moving funds from a Canadian to a US account or vice versa. With spreads of between 2% and 4% between the buying and selling prices, you can slowly get skinned alive if you are making frequent transfers and the brokerage firm (often a bank in the case of larger firms) will love you for it!

The Spread, the Market Order and the Limit Order

The spread is the difference between the bid and ask prices when you are placing an order to buy or sell shares. In some issues (such as GLD on the NYSE Arca) it is relatively small (less than 2%) and you would be advised to place a market order because it is so actively traded and you can go crazy changing your bids. In the case of some penny stocks, the spread can be quite large (20% or more).

If you place a market order, it will be executed at the ask price, if you are buying and at the bid price, if you are selling. However, it will also depend on how many shares are on offer at the time you place your order. I find it easier to illustrate this by way of an extreme example using a thinly traded penny stock.

Say you place a market order to buy 10,000 shares of XYZ and there are 1,000 offered at 10¢, 2,000 shares at 15¢ and 20,000 shares at 20¢. When you place a market order, you will get filled with the 1,000 shares offered at 10¢, 2,000 shares at 15¢ and 7,000 shares at 20¢. Your average acquisition cost will be 18¢ per share which is a far cry (80% more) from the ask price you saw for 1,000 offered at 10¢.

On the other hand, if you had placed a limit order for 10,000 shares of XYZ at 10¢, initially you would only pick up 1,000 shares; however, as the day wears on, you might pick up another 4,000 shares at the same price; leaving half your order unfilled.

Day Order and "Good Till" Time Limits

Continuing with the above example, if you had placed your order as a "day order" the order for the remaining shares would be cancelled at the close of trading that day.

Most self-administered online discount brokerage accounts will let you set a time limit on how long the order can be kept open. Usually this is 30 days, however, it may be longer in the case of some broker assisted accounts.

Lets say you put your order in "good for 30 days" (stating the actual the actual date) you will still have 5,000 shares of XYZ unfilled. Say that the next day it does not trade at all, your order remains unfilled but will be carried forward to the next day and so on until it is filled. Of course, someone else may come in and outbid you with a bid order of 100,000 shares at 11¢. If the shares are thinly traded, you may correctly assume that it is becoming unlikely that your remaining order for 5,000 shares will be filled at 10¢, so you may want to increase your bid price to 11.5¢ at which price you might be filled.

In this case, placing limits, while costing you extra commission (usually very modest on a self-administered online discount brokerage account) will save you a ton of money. If you work it out, your average cost (before commissions) would be 10.75¢ on these limit orders compared to the 18¢ had you placed a market order.

Watch Out for Commissions on Thinly Traded Stocks

If you are using an online brokerage account, you may be overjoyed at the low commissions on large trades. Thus, you could buy or sell $30,000 of the gold ETF GLD, and pay only $9.99 which is a minuscule 0.033%. On the other hand, if you purchase shares in a thinly traded penny stock and say, place an order for 30,000 shares at 10¢. If your entire order is filled on the one day, you would pay $300 for the shares plus a commission of $9.99 which is 3.33%.

If you were only able to pick up 15,000 on the first day and a similar amount of the second day, your commissions would be $9.99 per trade i.e., $19.98 or 6.6%. If you were only able to pick up 1,000 shares at 10¢, your commission would still be $9.99 on a $100 share purchase or 9.99% which is pretty steep.

Watch out for brokerage commissions on thinly traded stocks! They can add significants to your purchase or selling prices.

Settlement Date

When you purchase shares and your order is filled, you don't have to pay for them right away and the settlement date is three business days into the future. However, if you have a cash account, it is quite possible that the brokerage firm will want its money on the same day as the trade.

In the case of shares sold, you won't get your money until the third business day after the trade.

The Margin Account

A margin account lets you "borrow" from you broker, using the shares in your portfolio as collateral for the loan. I'll illustrate by way of a very simple example.

If you have $100,000 worth of good quality stocks in your portfolio, you might be able to get $50,000 on margin to purchase more shares. Of course the brokerage firm will charge you interest on the $50,000, if you use it.

I should point out that, depending on your broker, not all stocks are marginable. Thus some brokers will not consider shares of stocks that trade on the Toronto Venture Exchange; whereas they will accept those on the main TSX exchange. Others, place a dollar value limit such as $2.00 on the value of shares which can be accepted as marginable. Yet others have different levels of margin depending on the quality of the shares held in your portfolio. Even though some of this may sound confusing, your online brokerage account should show you how much money you have access to under margin.

Margin Calls

Margin calls are where things can get really nasty. I'll illustrate this point. Let's say you used the full $50,000 available on margin to purchase additional shares in the same company and the price of your original $100,000 in shares drops to $80,000; your revised margin at 50% will only be $40,000 yet you have bought $50,000 worth of stock.

This is where your broker can get very nasty. He will want an additional $10,000 right away to make up for the shortfall or he will sell shares in your portfolio to bring it back into balance and he is not going to be nice and see if he can get the best price for them. He will dump them. In a rapidly falling market, where the bids are thin, you will probably not get a very good price for what is sold.

Indeed, most margin calls come when the stock market is falling out of bed and it's not too difficult to get completely wiped out. So, be careful!

Brokers don't want to get caught with bad credit, so a margin account will almost certainly involve a credit check before being approved.

Short Accounts

If you are going to be involved in short selling, you will need to set up a separate account for these transactions. You will also need to have margin. I'll cover this in more detail in later.

Option Accounts

If you are gong to be involved in options, you will need to set up a separate account.

Retirement Accounts

Most brokers will allow you to set up a retirement account such as a RRSP account for Canadians or a 401(k) account for Americans. Under these arrangements, the funds and portfolios are kept segregated for income tax purposes.

Net Present Value (NPV)

Net Present Value (NPV) is frequently used when calculating the value of a mining stock.

In the case of a mining company, management is required, as part of their financial reporting obligations to shareholders, to state their reserves; usually reserves that have been clearly defined through drilling and those that are probable and inferred.

Once you know the reserves, you can calculate the mine life based on the current rate of production and in turn, you can calculate the cash flow up until the reserves are depleted and the mine ceases to operate (unless further extensions to the ore body can be found). Obviously, this requires some educated guess as to what future metal prices will be for the ore that is being mined.

Money in the future is not worth as much as it is today and there are always risks associated with mines such as flooding, mine cave-ins etc., so it is customary to apply some kind of discount to future cash flows. Then, add up the expected cash flows from future years and divide by the number of shares and you get the Net Present Value (NPV) per share. If no discount is used, it is referred to as a Zero Discount Net Present Value (ZDNVP).

The NPV or ZDNPV may be higher or lower than the current price of the shares. If it is higher, this would suggest that the shares are undervalued and conversely if it is lower, the price of the shares may be overvalued.

Mining analysts employ this technique all the time and you will find it frequently used in research reports. A quick and dirty way of doing the same thing is to calculate the total value of ounces of gold per share in reserves; multiple by some factor, such as the price received after mining and administrative expenses (say 50%) and compare that with the current share price. I've done this in the past to sift through a lot of data and it can be quite revealing; exposing companies that are overvalued and those that are undervalued.

EBITDA

EBITDA, pronounced "eh bee dah" stands for **E**arnings **B**efore **I**nterest, **T**axes, **D**epreciation and **A**mortisation and is a useful method of comparing large companies with significant assets or those with significant debt. Because it does not take debt repayments into account, it is not the same as cash flow and some critics claim that it can give a false impression of a company's financial health. Having said this, EBITDA could be used to see if a company is generating enough cash to make its debt repayments. Thus, you could divide EBITDA by the amount of debt repayments to arrive at a debt coverage ratio.

Some analysts use EBITDA to compare companies in the same industry.

Market Cap

Market cap (or "market capitalization") is the total value of a company's outstanding shares. It's easy to calculate. Simply multiply the number of shares by their current market price. It is the basis for the description of a company as being a "Large Cap" i.e., stock with a market capitalization over $10 billion; "Small Cap" with under $2 billion and "Mid Cap" in between.

From a fundamental analysis viewpoint, it has relatively little value, since there are better ways to calculate the value per share.

Net Book Value (NBV)

Book value is the value at which an asset is carried on a company's balance sheet and can be easily determined from the financials in a company's annual report by subtracting the cost of its assets minus the accumulated depreciation.

To calculate the Net Book Value (NBV), you would subtract goodwill, patents and other intangibles as well as sales taxes, service costs, etc. This is the value shareholders would receive in the event that the company was liquidated.

If you calculate the NBV on a per-share basis, you can compare it with the current price at which the shares are selling in order to determine whether they are overpriced or underpriced.

Price-to-Book Ratio (P/B Ratio)

The P/B Ratio (sometimes called the "Price to Equity Ratio") can be calculated by dividing the share price by the Book Value. It can also be used to determine whether shares are overpriced or undervalued.

Return on Equity (ROE)

The Return on Equity (ROE) is a useful metric to compare companies in the same industry. It is expressed as a percentage and is calculated by taking net income and dividing it by shareholder's equity. The net income in this calculation is before dividend payments to common shareholders but after dividends on preferred stock.

Debt/Equity Ratio

The Debt/Equity Ratio is a measure of a company's financial leverage. It is calculated by taking a company's total liabilities and dividing it by shareholders equity. When it is high (significantly above one), this should be cause for alarm, since the company could be exposed to increased debt carrying costs in times when interest rates are rising.

Company Guidance

Company guidance is a term used when management makes predictions about future revenue, earnings or performance.

I should caution you that not all companies make a habit of making such predictions but when they do, they can have positive or negative effects on share prices.

Be warned that some companies do a poor job of forecasting their future. A classic case was Research in Motion (RIM on the TSX now Blackberry or BB) which made poor sales forecasts and as a result lost a lot of investor confidence which is reflected in the very poor performance of its shares. On the other hand, I have noticed that Microsoft often underestimates its earnings. They may do this on purpose so that investors are pleasantly surprised when the actual results come out!

If you want to listen to company guidance calls, you can find out when they are scheduled by visiting *www.investorcalendar.com*.

Insider Trading

If XYZ company only has 10 million shares outstanding and I own one million of them, then I will own 10% of the company. Owning that high a percentage does not necessarily mean that I am a director or officer in the company, however, it would make me an "insider" because company management might consult with me about what's going on in the company. If I owned less than 10%, I would not be considered an insider in the US or

Canada, however, on the UK's AIM Exchange, I would be considered an insider if I owned over 3%.

Some investors use insider trading as an indicator of where share prices are heading.

Insider information can be obtained from the NASDAQ for stocks that are listed there. In the case of stocks in Canada, you can visit *www.Canadianinsider.com* to find out who is buying and who is selling and you can also purchase a full report.

Obviously, if insiders are buying, this suggests that they are on to a good thing and expect the shares of their company to increase in price. If they are selling, it could mean the opposite, however, it could also involve estate planning or personal financial considerations that have nothing to do with the company's future performance.

Management Compensation

Management compensation is usually set by a "compensation committee" made up of Board Members. I must say that I have no objection to good pay for good performance, however, this is not always the case.

In many instances, compensation for poor performance can border on being obscene but there is little the common shareholder can do other than turf the Board out at the next annual meeting (which rarely happens because a cosy relationship may exist between the Board and major shareholders). When you uncover such abuses, in practice, there is little you can do other than dump your shares and find another company where the shareholders are not being raped by management.

Institutional Holdings

If the company you are interested in is listed on the NASDAQ, you can visit their Web site to find out what percentage of the shares are held by institutional investors. The site will also tell you how many positions are new, how many sold out and the increase or decrease in the positions of those who already own the stock.

From your viewpoint as an investor, my advice would be to be careful about buying stocks with no institutional holdings because they probably don't meet the investment criteria that they have established. On the other hand, you don't want to own shares in companies that have a very high institutional ownership, because you may get crushed in an exit stampede in the event that they all decide to sell.

Corporate Web Sites, Annual Reports, Annual Meetings and Conference Calls

Practically every listed public company has a Web site which can be a trove of information about what the company does, its management, Board of Directors, news releases, etc.

If you are interested in a particular company, you can talk to their investor relations department and you can have them mail you copies of their annual and quarterly reports. Most of these are also available in PDF format and can be downloaded online.

Even if you can't actually attend an annual meeting, some companies provide live streaming video of the event which may be archived so that you can access it at a later time from their Web site.

Some listed companies hold conference calls over the phone for analysts and in many instances there is nothing to prevent you from listening in. Again, some are archived so that you can listen to them after the event.

Analysts' Reports

Stock market analysts generally specialize in an industry group and may only report on a few companies within that group. If you are using a full service broker at a large firm, you should be able to obtain copies of reports from their research department. In addition to that, some companies such as Morningstar, Standard and Poors plus Thompson Reuters provide paid access to reports by analysts.

Watch Out for Market Expectations!

Based on analysts' reports and company guidance, the market builds up expectations of what's going to be announced; especially annual and quarterly earnings. In many cases, even though earnings might have improved on a year-over-year-basis, they might fall short of analysts' expectations and the shares take a tumble. If they exceed expectations, the shares usually rise.

On other occasions, the market has come to anticipate that a company will make a "good news" announcement such as in the case of a mining company that is expected to release recent drill results or pre-feasibility study reserve figures. When the actual announcement is made; even though it may be good, the market sells off because investors who have participated in the run up want to take profits or those who have been waiting patiently to sell, decide that it's time to exit.

The reverse may happen, when bad news is expected and the shares have fallen in price. The actual announcement may "clear the air" and the shares could rise in price!

On the other hand, when bad news is expected and is released, the shares may continue their downward path and the opposite when good news is released.

Tricky–isn't it? Just be careful how you play market expectations.

Chapter 14

Short Covering Gold Rallies

What is Going Short?

When you buy shares in a company, technically you are "going long" which means that you would hold the shares for a period of time and eventually sell them.

On the other hand, you could sell shares in a company without ever owning them in a process known as "going short." In this scenario, you would borrow the shares from an investor who owns them. This borrowing is done through the facilities of your stock broker who is holding shares owned by other investors. Once you have borrowed the shares, you can sell them. At some point in the future, you will have to buy the shares back and return them (via your broker) to the person who loaned them to you.

The whole point of going short is that you are hoping that the share price will fall so that you can buy them back at a lower price than that at which you sold them.

Unlimited Risk

Don't get carried away and start shorting everything in sight, because you are exposing yourself to almost unlimited risk. If you had shorted Detour Gold

Mines in August 2013 and covered your short position in late November of the same year, you would have almost tripled your money. However, if you had gone short in late November 2013 and covered in mid-March 2014, you would have lost almost two-thirds of your money.

Liable for Any Dividends Payable

Since you have borrowed the shares when you go short, you are responsible for paying any dividends back to the original owner during the time that you have borrowed them.

ETFs that Let You Go Short

The challenges of going short are that you have to borrow the shares; you are responsible for paying any dividends during the time you have borrowed the shares; the shares may be called at any time and you could get trapped in a short squeeze which forces you to cover; regardless of how high a price you have to pay.

Rah!..Rah!...trumpet roll–enter the ETF that let's you go short and avoid most of these problems since your risk is spread out. For example, you could go short gold stocks or gold as a commodity.

Just remember one thing. When you go short, you don't want to do so for ever. There is an overall tendency for markets to increase in value over time; even though they can take some weird downturns along the way. So, at some point or other, you'll have to cover your position i.e., sell the ETF that is a short. In my opinion, it's best to use short ETFs as a limited-time trading strategy--get in, take your profits (assuming you have some) and get out!

Are Shares Available to be Borrowed?

While it rarely happens with widely traded issues, I've certainly run into a situation where the dealer has been unable to borrow the shares and consequently it was not possible to enter into a short position.

Call Options

A Call Option is the opposite to a put option. It is a contract that gives the owner of the call the right to buy a specific number of shares of a certain company within a specified time. I should point out that the purchaser of the call is not **obliged** to buy. Calls are particularly useful when shares of a company are rising. Thus, if you purchase a call for 100 shares of XYZ company at $100.00 and the price rises to $130.00, you can exercise your call. You would short sell 100 shares at $130.00 and cover your short position by purchasing from the option writer at $100.00; thereby making a profit of $30.00 per share times 100 shares ($3,000) less the cost of the premium to purchase the call.

If Shares are "Called"

This is different from what I just discussed and does not involve options.

It is unlikely to happen when a company has a large number of shares issued and they are not concentrated in a few hands. On the other hand, if you get down into the penny stock area and if there are relatively few shares issued (say 10 million) and many of these are tightly held by management and insiders then you could run into a situation, if you have borrowed shares and the person you borrowed them from wants them back. This is were the shares are "called" and you may end up scrambling to cover your position at ever and ever higher prices.

Illegal? No, for the most part it's fair game and it's just your bad luck if you got caught in the squeeze.

Short Covering Rallies and a "Short Squeeze"

When you read the financial press, you will often see reference to short covering rallies. This occurs when the shares of a company run up in price causing those who are short to cover their positions; either to lock in their profits or to avoid creating a loss. Such action is also referred to as a "short squeeze." This type of thing also occurs in the commodity markets and gold prices have been subject to some spectacular run-ups when the shorts are caught off-guard by some unexpected event. These types of sharp rallies were quite common in 2013 but I have noticed that they seem to be much more muted in 2014.

Chapter 15

Mutual Funds Which Specialize in Gold

Mutual Funds

Basically, a mutual fund pools the money from a number of investors and then uses those funds to purchase a collection of stocks, bonds or other assets. They are managed by a fund manager who makes investments according to the fund's objectives. You can buy into them by purchasing units, either through a stock broker, discount broker, investment advisor, bank or directly from the fund itself.

They became very popular about 20 to 30 years ago but that popularity has been waning a little in recent years since Exchange Traded Funds (ETFs) can offer the same kinds of exposure with a lot less hassle. An interesting article which appeared on December 17, 2011 in Canada's *Financial Times*, showed that so far as Canadians are concerned, they have twenty times more of their money invested in mutual funds (about $0.75 trillion) as they do in ETFs ($42.4 billion).

Of course, mutual funds have a tremendous benefit to their market dominance over ETFs, since they are advertised and marketed. The cynic in me also tells me to point out to warn you that there is more money in fees to be made by fund managers and the people who sell you the fund than there is in buying an ETF through an online discount brokerage account!

Mutual Funds that Invest in Gold

There are a couple of gold mutual funds which may be of interest to you. The Millenium Bullion Fund invests in gold, silver and platinum. It purchases the physical metal and does not play the futures market.

AGF Management, BMO, Franklin Templeton, Gabelli, First Eagle and Invesco all operate gold and precious metal funds. In addition, Fidelity Select operates a gold portfolio fund.

Open and Closed-End Funds

An open-end fund is one whose units you can purchase and sell with ease based on the value of the underlying assets in the fund. If you have purchased units in an open-fund that then becomes "closed" you will only be able to sell them to another buyer and not back to the fund itself. This will often mean that you will lose money on your investment since the "other purchaser" may not be willing to pay for the value of the underlying assets in the fund.

Management Expense Ratios (MERs) and Commission Fees

You can expect to pay management expense fees of anywhere from about 0.5% to 1.0% for an index fund and up to 3% for a fully managed fund. These management fees are deducted from the assets in the fund and comparisons can be made with other mutual funds by dividing its operating expenses by the average dollar value of the funds under management. This figure is referred to as the Management Expense Ratio or "MER."

In addition, there are sales commissions payable to the agent who sold you the fund. I was going to say that these come in three different "flavours," but I'll use the words "bitter pills" as being more appropriate! In some funds, the fees are collected up-front, when you purchase. These are called "front-load" fees and generally run around 2% however, you can sell your holding at any time without further charges.

The second "bitter pill" is a redemption charge when the fund is sold. These can be very bitter and run at over 5% during the first few years you hold the fund but usually decline thereafter to 0% after seven or more years. I don't apologize for my cynicism, but they are designed to handcuff you to the fund.

The third "bitter pill" comes with a sweet sounding name: "no-load" however, in practice there is usually a load, in the form of higher management fees. Deceptive?

As if that wasn't enough, many funds also have trailer commission fees of up to 1% payable to the party which sold you the fund! These fees are paid annually and taken out of the assets in the fund, and you, as an investor can't see them because they are hidden!

As you may gather from my comments, I'm not a great fan of mutual funds!

Mutual Funds vs. ETFs

Both mutual funds and index ETFs allow investors with a relatively small amount of money to get exposure to a wide selection of companies which would be difficult for them to achieve on their own. This can be particularly useful if you are a beginner with only a small amount of money to invest.

As I pointed out previously, you can purchase an ETF and trade it like a stock whenever the markets are open (and even after hours in some cases). In the case of a mutual fund, you can only purchase it based on its value at the close of markets; once its "value" for that day has been established.

One disadvantage of owning mutual funds are the commissions which I covered. On the other hand, when you purchase an ETF all you are paying is one stock brokerage commission. However, a management fee is also charged by the ETF manager which is subtracted from the value of the holdings and is spelled out in the prospectus.

Probably the worst feature of mutual funds is the selling pressure you are under from fund managers and the dealers, who want to make commissions, and that is something which you will avoid when you purchase ETFs.

The advantage of ETFs is that you can easily re-balance your portfolio.

Further Information on Mutual Funds and ETFs

A leading source for information on mutual funds and ETFs is Morningstar. Much of the analysis is free and can be accessed at: *www.morningstar.com*.

Chapter 16

Investing in Gold Coins, Bars and Wafers

The Premium

You will always have to pay a premium over the daily gold price whenever you buy coins, bars or wafers. Since the premium can vary quite considerably, I would recommend that you pay close attention to it.

Gold Coins for Investors

I would divide gold coins into two categories: coins for investors and coins for collectors. I'll cover the former in this section.

The one ounce Krugerrand was first minted in South Africa in 1967. It is 91.7% gold, with the balance made up of copper which gives it a rich orange colour. It was never issued as "currency" in South Africa but only to be sold for its value in gold.

By way of interest, I compared the price of gold to one dealer's buy and sell prices on May 23, 2014. The results were as follows:

The dealer's sell price for 10 Krugerrands: $1,399.10 each.

The New York NYMEX price of gold at close on May 23, 2014: $1,293.90 per ounce.

The dealer's buy price for a single Krugerrand: $1,293.90.

On the dealer's sell side, you are paying a premium of 8.1% and on the dealer's buy side, you are getting a discount of 0.0%. The spread between the dealer's bid and ask is 8.1%.

When looking at these comparisons, bear in mind that the Krugerrand is not pure gold, so you'll have to adjust the figures for the true gold content. I've done the calculation as follows:

The dealer's sell price based on gold content of 0.917 oz: $1,282.97 each.

The New York NYMEX price of gold at 11.16 a.m.: $1,186.5 per 0.917 ounce.

The dealer's buy price based on gold content of 0.917 oz: $1,186.51.

In this case, you'll see that both the dealer's buy price is virtually the same as the comparable amount of metal but the sell prices are higher by 8.1%.

The Krugerrand gained great popularity after its introduction, which led to copycat coins being issued by other nations. For example, the United States minted the Gold Eagle in 1986 and it is available as a one-ounce coin, a half-ounce, a quarter-ounce and a one-tenth ounce. The gold content is similar to the Krugerrand.

Canada minted the Maple Leaf gold coin series in 1979. These are also available as a one-ounce coin, a half-ounce, a quarter-ounce and a one-tenth ounce. They are made of pure gold and this certainly added to their appeal.

Old Gold Coins

Once you start to get into old gold coins you have to take the numismatic value into account. This is a field for collectors and specialists, so tread carefully, because the premiums for gold content can be very large!

Gold Bars

Gold bars are generally sold in one kilogram weight; equivalent to 32.1507 troy ounces. Unlike coins for investors, the purchase of a kilogram gold bar can be quite expensive and could set you back $42,848.89 at the time of writing. The premium was only $14.99 or 0.035% above the actual price of bullion. As you can see, the premium is very small.

Gold Wafers

If the cost of a kilo gold bar will take up all your housekeeping money for the next ten years, you can purchase pure gold wafers ranging in weight all the way from half a kilogram down to a quarter ounce. The problem is that as you go down in size, the premium increases.

Bullion Dealers

There are many respectable bullion dealers who will sell you gold coins or bars. They will sell over-the-counter or ship to your door. Just watch out for the "handling charges". I've been told that they can be as high as 15% at some banks.

Chapter 17

The Gold Futures Market

Playing the Futures Market

Instead of buying physical gold, gold mining shares, ETFs or mutual funds, you can purchase gold futures on the COMEX, which I referred to earlier in this book. Indeed, twenty times more gold is traded on the COMEX than is mined by gold mining companies each year! However, you don't have to buy enormous quantities in order to play the game. Thus, the standard contract is 100 troy ounces which at the spot price on May 23, 2014 would be about $129,290. Relax! You can buy futures on margin, but this is where it gets scary!

The margin requirements vary according to the dealer who is trading for you, but the minimums are also set by the futures exchanges themselves. In fact, with the frothy market initial margin requirements were increased on August 2011 to $11,475 and then reduced several times. On May 23, 2014, they stood at $6,600 per initial contract.

COMEX isn't the only major gold futures exchange. Another is TOCOM in Tokyo. In their case, the contract size is smaller than that of COMEX and is set at 1 kilogram which makes it less than one-third of the size of the COMEX contract.

If you want to get quotes for gold futures, you can visit *www.tradingcharts.com*. There, you can find 10-minute delayed gold future quotes on the COMEX Gold (Globex) Futures Exchange. The further you go into the future (in general), the greater will be the price.

Let's take a look at the futures prices after the close on May 23, 2014 which were as follows:

Spot price for June 2014	\$1,293.40
August 2014	\$1,288.30
October 2014	\$1,216.70
December 2014	\$1,217.50
February 2015	\$1,242.70
June 2015	\$1,220.70
December 2015	\$1,336.30
December 2016	\$1,793.00

So, if you are going to purchase December 2014 gold, you would put up \$6,600 for the 100 troy ounces valued at \$121,750 or approximately 5.4%. Since prices are constantly fluctuating, you may get a call to put up more money to maintain your position. This can happen on a daily basis and it is the reason why many contracts are only held for a day. Note that you'll also have to pick up the purchase commission as well as the commission when you sell.

On the day I recorded these quotations, the spot price was \$1,292.90 so the futures market anticipates that gold prices will remain fairly sluggish until the end of 2015 but will start to perk up shortly thereafter.

As in the stock market, you can go short and you can also place stop loss orders. Be warned, however, that dealers are empowered to close out losing positions if you don't immediately put up more money when you receive a margin call.

Just like the stock market, you are either purchasing from another unknown individual or selling to an unknown individual and the futures trader is acting as the intermediary or "broker." As with trading stocks, you can employ a full service futures broker or trade online.

Normally, you would sell out your position before you would be required to take physical delivery of the gold. If you judged the market right, you would pocket a profit and if not, you would take a loss. If you don't sell, you will have to put up the full amount of money for the contract, plus delivery charges and arrange delivery to a vault for safekeeping.

Chapter 18

Conclusion: The Reality Check!

The Sell-Off in Gold ETFs Seems to Be Abating

According to figures compiled by GFMS, Thompson Reuters, The London Gold Fixing and the World Gold Council, in the first Quarter of 2014, the outflows of gold from ETFs was zero compared with 177 metric tons for the corresponding period in 2013.

Investors Shifted from Gold Into Hot Equities

On of the major contributing factors to the sell-off in gold ETFs and gold company shares has been the rush into the hot equity market during 2013. Why stay stuck in gold when you can purchase high-flyers like NetFlix, Amazon and junior biotech companies? Well, that may be coming to an end, as some of the high fliers had their wings severely clipped in May 2014. Much of this sell-off has gone unnoticed because most of the companies involved were not major movers insofar as the stock indexes were concerned.

So long as investors can be titillated by the prospects of surging equity prices and the possible recovery of some of the high-flying darlings which dropped of the perch, gold prices could remain under pressure. However, I must say that I get very nervous when I see vast numbers of investors charging into a hot stock market and I can't help feeling that a correction is long-overdue. The question is: will it be severe enough to cause investors to move into gold as a safe-haven or will it be mild enough to stop gold falling and assume an "up-down" sideways dance?

Becoming too focussed on changing asset allocations in North America may be myopic and cause investors to miss out on the big gold picture, because there are other more important factors at play.

China Will Continue to Accumulate Gold

Even though the Chinese Central Bank's holdings of gold have not budged in the last year or so, it appears that its sovereign wealth fund is stealthily accumulating significant amounts of the metal. In addition, the emerging middle class is using its wealth to purchase jewellery, especially to celebrate the Chinese New Year and Valentine's Day to such an extent that the country is a major importer of the metal, since its internal mine production falls well short of its needs.

So far as I can see, the only impediment is if the economy falters which could put the currency under pressure, but even that could result in increased gold purchases to offset the currency's decline.

Also, if China's aggressions with its neighbours in the South China Sea should start open conflict, there could be a rush to purchase gold as a safe haven.

India Could Be a Game Changer

As I mentioned earlier, Hindu nationalist, Narendra Modi's MJP party swept to power in the May 2014 elections. There are also hopes that his government will loosen up on the import duties and regulations governing gold and, if that happens, it could be a major game changer for a country that has traditionally been one of the world's largest consumers of gold.

The Debt and Currency Crisis Will Continue

In mid-May 2014, the European debt crisis seemed to be contained, however, there were signs that not all is well. The European parliamentary elections in May saw the election of many members who are opposed to the EU and would like to see it broken up.

In addition, many national governments are swinging to the right with anti-immigrant sentiments and a desire to exit the Union.

As I mentioned earlier, The European Central Bank is wrestling with a bond purchasing program to try to purchase bonds of the southern peripheral nations like Greece, Spain, Portugal and Italy. Such a "quantitive easing" program is under intense criticism from many German politicians who don't see why Germany should help economies that are floundering.

All of this bodes ill for the cohesion of the Eurozone which is further exacerbated by independence movements in many EU member nations, such as Scotland, Spain's Catalonia and Belgium's Flanders.

Needless to say, all this commotion and uncertainty is likely to benefit the price of gold.

Inflation is Likely to Return

In mid-May 2014, the "official" rates of inflation in the US, Canada, the UK and the EU seemed relatively benign but, as I said earlier, the "real" rates of inflation, if fuel and food are taken into consideration, far exceed the official figures and sooner or later these will work their way into higher consumer prices as increased costs will have to be passed along.

Even though many "pundits" think that gold prices will benefit from inflation, I think that gold will only be slightly influenced, unless we head into hyperinflation, which seems to be an unlikely scenario at the time of writing.

Gold Will Continue to be a Safe Haven in the Face of Geopolitical Uncertainties and Terrorism

Unfortunately, world peace is unlikely to break out anytime soon! Gold benefits from the "fear trade."

As the Ukraine crisis demonstrated, gold will react positively to any major geopolitical crisis. "Tzar Vladimir" wants his legacy to herald a reversal of the breakup of the USSR regardless of the economic cost to the average Russian. In addition, his ambitions for dominance of Arctic natural gas and oil deposits will throw Russia into conflict with other nations which have interests in that area such as Canada, the US, Norway and Greenland. Unfortunately, his antics are likely to have a positive effect on the price of gold.

China's territorial claims in the South China Sea over oil and natural gas exploration rights and fishing rights are already causing military tensions with neighbours such as Vietnam, Indonesia, the Phillippines and Japan. It only takes a few missteps and a major conflagration could ensue.

Add to all of this the nuclear threat from Iran, which in my opinion is playing for time with its de-nuclearization talks, while it clandestinely develops a bomb and the rocketry to deliver it. Also, there is some 20-plus year old kid in North Korea with his finger on the nuclear button. None of this bodes well for de-escalation, but unfortunately, they are also positive for the price of gold.

You Be The Judge!

No, I'm not going to put my neck on the line and predict the price of gold at the end of 2014 or its price at the end of 2024. The bottom line is that nobody knows, and only fools tell you they know for sure!

What I have tried to do in this book is to alert you to some of the factors which influence the price of gold. Hopefully, it has been a reality check. At the end of the day, you'll have to be the judge!

PRODUCTIVE PUBLICATIONS

Our 29th Year Publishing Non-Fiction Books to Help You Meet the Challenges of the Digital Age

Other Great Books to Help You! Check Them Out!

ou will find brief outlines in the following pages but for more details and chapter ontents, visit our Web sites as follows:
SA and Overseas: *www.ProductivePublications.com*
anada: *www.ProductivePublications.ca*

ou can order using a credit card as follows:
rder Desk USA and Canada Phone Toll-Free: 1-(877) 879-2669
ax Order Form on the last page to: (416) 322-7434
nquiries Phone: (416) 483-0634
lace Secure Orders Online at:
SA and Overseas: *www.ProductivePublications.com*
anada: *www.ProductivePublications.ca*

-mail orders to: *ProductivePublications@rogers.com*

Snail Mail Orders to: Productive Publications
P. O. Box 7200, Station A
Toronto, ON M5W 1X8 Canada

ourier Orders to: Productive Publications
7-B Pleasant Blvd., #1210

coin Mania

e Birth of a Worldwide
tual Currency or the
art of an Insane Bubble?

: Learn2succeed.com
orporated

delightfully cynical book
ch looks at the fiat paper currencies which we use everyday
questions how long they can survive under the crushing
ght of mountains of national debt. Could they be replaced
a digital cryptocurrency called Bitcoin which threatens to
rturn the tables of traditional fiat money changers?

pages; Softcover; ISBN: 978-1-55270-500-1 CIP
ada: $24.95 US: $24.95 UK: $15.79

Crowdfunding for
Beginners:

How to Raise Money for
Start-Ups, Early Stage
Enterprises, Charities and
Non-Profits

By: Learn2succeed.com
Incorporated

More than $2.7 billion is being raised every year through crowdfunding. Most of the money is raised by charities and non-profits; as reward-based fundraising for books, theatre productions and lending for personal debt. Equity-crowdfunding is gaining traction in the UK, Australia and New Zealand but is struggling under a regulatory burden in the US and Canada.

156 pages, Softcover, ISBN 978-155270-706-7 CIP
Canada: $24.95 US: $24.95 UK £15.79

f-Publishing for
ginners:

v to Prepare, Edit,
olish and Market Your
n Print Books and
ooks

Learn2succeed.com
orporated

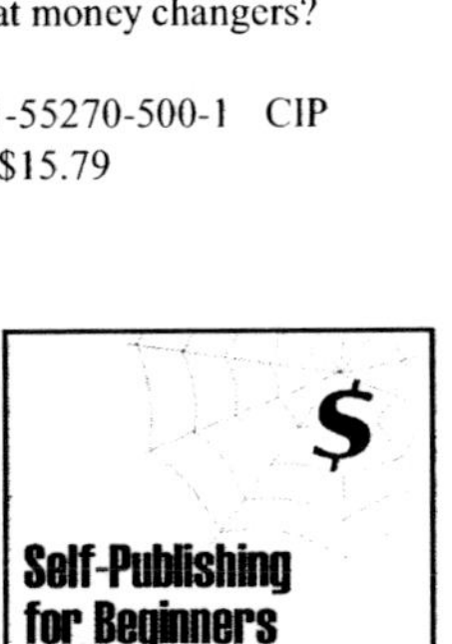

nputer software to help
are a manuscript and conduct research. Different ways
dit and proofread and all the cataloguing, copyright and
stuff. How to publish print books and advertise,
ket and sell them online. How to prepare eBooks; the
rent formats; how to handle covers and some of the
publishing alternatives. Learn about pricing and how to
eBooks through resellers.

pages, Softcover, ISBN 978-155270-488-2 CIP
ada: $24.95 US: $24.95 UK £15.79

Mountains of Sugar,
Canyons of Salt,
Pools of Fat:

Welcome to Your Local
Grocery Store!

By: Learn2succeed.com
Incorporated

This book lambasts food producers and retailers for not doing more to help prevent diabetes, heart disease and strokes which cause huge numbers of deaths. It examines the main culprits: salt, fat and sugar. It looks at prepared foods, growth hormones and genetically modified foods. It criticizes the grocery trade for doing pathetically little to encourage customers to purchase healthy foods that are good for their health.

58 pages, Softcover, ISBN 978-155270-712-8 CIP
Canada: $14.95 US: $14.95 UK £8.95

For details visit our Canadian Web site: *www.ProductivePublications.ca*
American Web site: *www.ProductivePublications.com*
Order securely online or mail the order form at the end of this catalogue
Phone our Order Desk toll-free at: 1-(877) 879-2669

Be Your Own Boss: Start a Business!

How to Hit the Ground Running

By: Learn2succeed.com Incorporated

Where to get an idea for starting a business and find out if it can make money. Learn about the legal forms of business; how to set objectives and begin operations. Computer hardware and software that can help you get off to a flying start. The basics of advertising; both offline and online. Inexpensive and easy ways to create a Web site and sell products or services over the Internet. How write a business plan.

156 pages, Softcover, ISBN 978-155270-709-8 CIP
Canada: $24.95 US: $24.95 UK £15.79

Make Money Selling Your Digital Products Online:

How to Create, Advertise and Market eBooks, Streaming Videos and Audio Books

By: Learn2succeed.com Incorporated

Create, sell and deliver digital products over the Internet. Open a Web site and sell online. Ways to promote your site and improve your chances of being discovered. How to use metadata to help in the "discoverability" of your individual titles.

148 pages, Softcover, ISBN 978-155270-715-9 CIP
Canada: $24.95 US: $24.95 UK £15.79

Business Start-up n the Digital Age:

Smart Ways to a Successful Launch

By: Learn2succeed.com Incorporated

Helps inventors a well as those who want to sell an existing product or find something appropriate. Start a service business. Selecting a name and legal structure. Home-based business versus commercial an retail leases. Calculate human resource requirements. Onlin and offline advertising. Set up a Web site to sell online. Calculate the money needed to start. Software for creating a business plan.

164 pages, Softcover, ISBN 978-155270-522-3 CIP
Canada: $24.95 US: $24.95 UK £15.79

A Raging Gold Bull or a Hibernating Golden Bear?

Investment Reality Check

By: Learn2succeed.com Incorporated

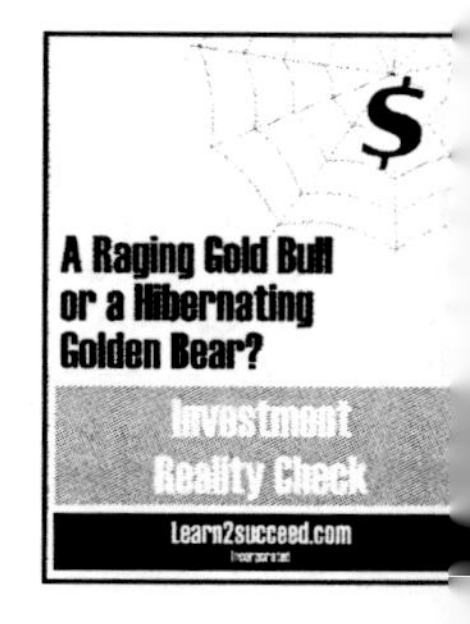

Written by a former mining analyst, under a pseudonym, this book covers some of the less publicized factors which influence the price. These will be of interest to gold bugs as well as those who are curious about investing in the yellow metal.

132 pages, Softcover, ISBN 978-155270-703-6 CIP
Canada: $19.95 US: $19.95 UK £12.95

siness Start-Up
Beginners

w to Become
ur Own Boss

: Learn2succeed.com
orporated

ou have ever dreamed of starting your own business and oming your own boss, you've taken the first important step selecting this book. It will show you what it takes to ome an entrepreneur and how to find ideas to start your 1 business. You will have to acquire a lot of skills very ckly and this book will alert you to some of the things you need to know and provide you with a lot of insight, based irst-hand experience.

pages, Softcover; ISBN 978-1-55270-444-8; CIP
ada: $19.99 US: $19.99 UK: $12.59

nture Capital Financing
Beginners

w to Raise Equity
ital from Venture
italists and Angels

Learn2succeed.com
orporated

u are already in business or about to start a business, book will help you raise equity capital from traditional ure capitalists or from "angels".

pages; Softcover, ISBN: 978-1-55270-458-5 CIP
ada: $19.99 US: $19.99 UK: $12.59

Bank Financing
for Beginners

How to Borrow Money to
Grow Your Business

By: Learn2succeed.com
Incorporated

This book should definitely be on your reading list before you go charging into you bank in search of a business loan. Credit conditions may be tight, but this book should increase your chances of securing a bank loan.

114 pages; Softcover; ISBN: 978-1-55270-459-2 CIP
Canada: $19.99 US: $19.99 UK: $12.59

Public Speaking
for Beginners

How to Communicate
Effectively in the
Digital Age

By: Learn2succeed.com
Incorporated

The King's Speech, drew attention to public speaking by someone with a disability. Even if you don't have a disorder, speaking in public can still be a challenge. This book will help you communicate effectively with your audience.

78 pages; Softcover; ISBN: 978-1-55270-452-3 CIP
Canada: $15.99 US: $15.99 UK: £9.99

For details visit our Canadian Web site: *www.ProductivePublications.ca*
American Web site: *www.ProductivePublications.com*
Order securely online or mail the order form at the end of this catalogue
Phone our Order Desk toll-free at: 1-(877) 879-2669

Your Guide to Raising Venture Capital for Your Own Business in Canada

Revised and Updated 2013-2014 Edition

By: Iain Williamson

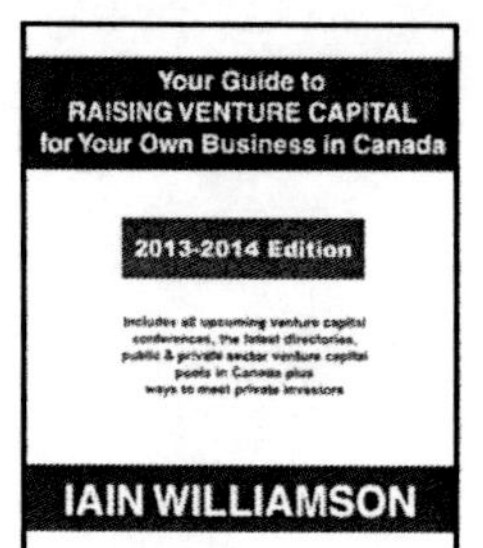

This book is a gold mine of information if you are raising venture capital in Canada. It shows you how to do it yourself. Lean about the structure of the industry; what venture capitalists are looking for and how they evaluate deals. You are given tips on the negotiating with them. It lists 86 conferences and provides 544 contact addresses. Find out what angel investors are looking for and how they could help you.

248 pages, softcover; ISBN 978-1-55270-667-1 CIP
ISSN 1191-0534 Canada: $74.95

Your Guide to Arranging Bank & Debt Financing for Your Own Business in Canada

Revised and Updated 2013-2014 Edition

By: Iain Williamson

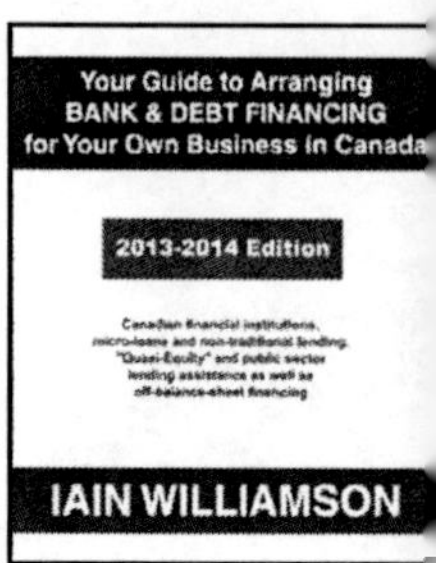

Learn the secrets of successful debt financing in Canada. Fin out who the players are. Covers 373 loan programs an includes 843 contact addresses. How to prepare your compan before you approach lenders. Find out how your loa application is evaluated. Can factoring or leasing help you The author has many years of experience in bank financing an leasing and his book will help you in your quest for a loan.

386 pages, softcover; ISBN 978-1-55270-668-8 CIP
ISSN 1191-0542 Canada: $81.95

Your Guide to Financing Business Growth by Selling a Piece of the Pie

What's involved in going public; employee share ownership plans and franchising in Canada

Revised and Updated 2013-2014 Edition

By: Iain Williamson

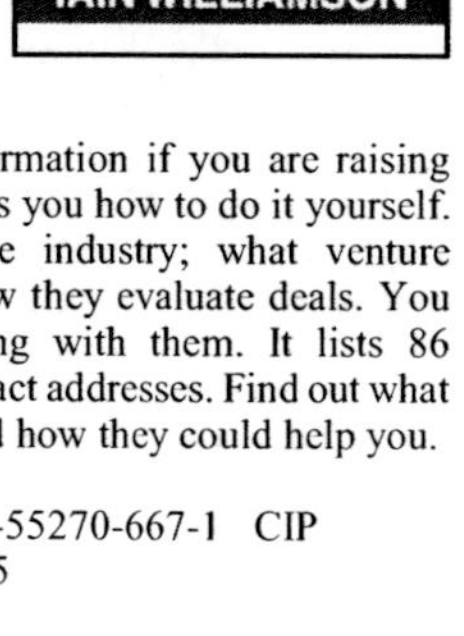

A critical examination of three methods of growing your business by using other peoples' money. How to sell shares to the public or to your employees. Covers the Canadian stock exchanges and their listing requirements. Includes sections on London's Alternative Investing Market (AIM) and on NASDAQ in the US. What's involved in establishing an Employee Share Ownership Plan (ESOP). How to expand through franchising. The author was a financial analyst in the Canadian stockbrokerage business for five years.

162 pages, softcover; ISBN 978-1-55270-669-5 CIP
ISSN 1191-0488: Canada: $46.95

Your Guide to Canadian Export Financing: Successful Techniques for Financing Your Exports from Canada

Revised 2013-2014 Edition

By: Iain Williamson

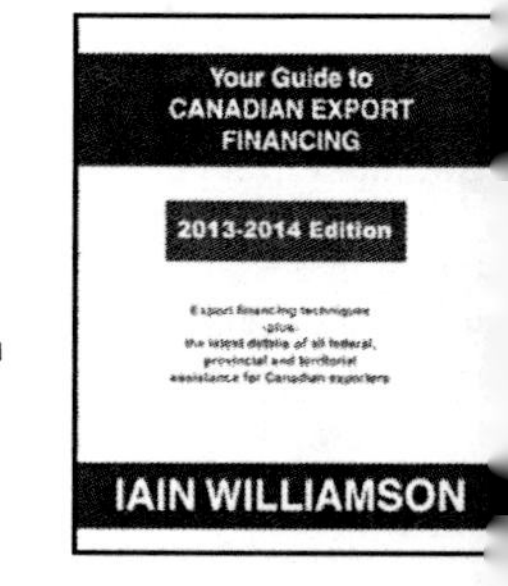

Provides you with practical techniques for financing yo exports. Get details of all provincial, territorial and feder assistance programs that help you export including address and phone numbers to steer you in the right direction. Includ a chapter on insurance and sources of marketing informati for exporters. The author is a consultant and entrepreneur wh knows the practical side of importing and exporting.

300 pages, softcover; ISBN 978-1-55270-670-1 CIP
ISSN: 1191-047X Canada: $58.95

r Guide to Starting elf-Financing r Own Business Canada

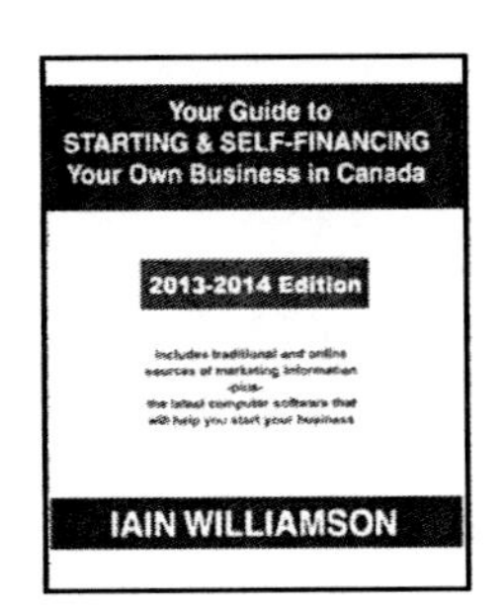

ised 2013-2014 Edition

Iain Williamson

2013-2014 Edition has updated to reflect the many ges that have taken place in sources of marketing information. Shows you how to ate a business out of your home. How to use computers the Web to run your business more efficiently. Covers web oring software and how to sell online with your own e-merce site or through eBay; as an Amazon Merchant or ugh online classifieds. Helps you determine how much ey you really need and whether you can self-finance your business to compete in the digital age.

pages, softcover; ISBN 978-1-55270-665-7 CIP
N 1191-0518 Canada: $56.95

r Guide to vernment ancial Assistance Business

parate Editions-one for h Province & Territory)

ised 2013-2014 tions

Iain Williamson

ness financing in Canada is in a constant state of flux. government programs are continually being introduced. ones are often amended or discontinued with little icity. These books will provide you with the latest mation on all Federal and Provincial/Territorial programs specifically relate to each area. Author, Iain Williamson, ntrepreneurial Business Consultants of Canada, has over ears experience as a stock market financial analyst and as er-manager of his own companies.

95 ea. Softcover; CIP. Title & ISBN See list on right →

Your Guide to Preparing a Plan to Raise Money for Your Own Business

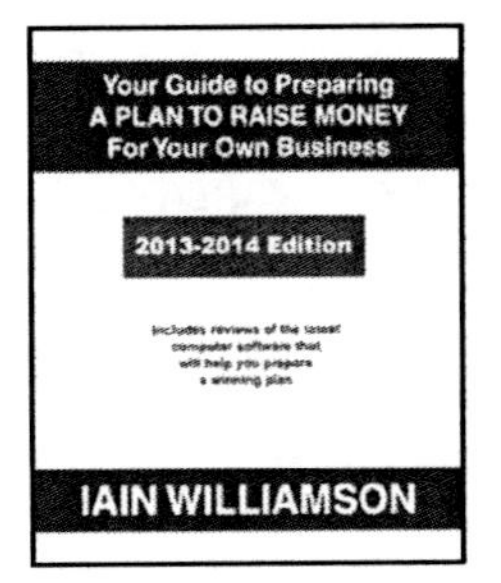

Revised 2013-2014 Edition

By: Iain Williamson

A good business plan is essential to succeed in your quest for financing. Contains a step-by-step guide to create your own winning plan. Computer software you can use. Find out how spreadsheets can help you. Learn how to address the concerns of investors or lenders. Tips on structuring your plan. Contains a sample plan as an example. Computer software to help you make great presentations to investors or lenders.

310 pages, softcover, ISBN 978-1-55270-666-4 CIP
ISSN 1191-0496 Canada: $46.95

Your Guide to Government Financial Assistance for Business In...

EDITION	ISBN	PAGES
Newfoundland & Labrador	9781552706718	304
Prince Edward Island	9781552706725	273
Nova Scotia	9781552706732	308
New Brunswick	9781552706749	278
Quebec	9781552706756	312
Ontario	9781552706763	376
Manitoba	9781552706770	314
Saskatchewan	9781552706787	306
Alberta	9781552706794	306
British Columbia	9781552706800	286
The Yukon	9781552706817	256
The Northwest Territories	9781552706824	246
The Nunavut	9781552706831	248

Please specify Province or Territory when ordering. All titles are $89.95 each.

For details visit our Canadian Web site: *www.ProductivePublications.ca*
American Web site: *www.ProductivePublications.com*
Order securely online or mail the order form at the end of this catalogue
Phone our Order Desk toll-free at: 1-(877) 879-2669

Steps to Opening a Successful Web Store

The Basics of How to Set-Up Shop in Cyberspace

By: Learn2succeed.com Incorporated

How to find products to sell, research the market and open your own e-commerce Web site. Use online and offline advertising to drive traffic. Consider selling at eBay auctions at your own eBay store. Review the advantages of running your business from home and how to set it up. Learn how to select the best computer hardware and software. Establish your marketing strategy, prepare your marketing plan and incorporate it into your business plan. Calculate how much will it cost to start.

154 pages, softcover; ISBN 978-1-55270-357-1 CIP
Canada: $24.95 US: $24.95 UK: £15.79

Steps to Starting a Successful Retail Business

How to Find a Niche and Turn it Into a Money Machine

By: Learn2succeed.com Incorporated

Where to get ideas and direct import your own merchandis How to research your market. Covers legal structures and ta permits. The importance of location and the hazards of leas renewals. Your store layout, fixturing. signage, lighting an window displays. Hire staff and reduce shoplifting. Us planogramming to fine-tune your operation. How to advertis and use direct marketing to augment your sales. Use commerce on the Web to extend your reach.

144 pages, softcover; ISBN 978-1-55270-359-5 CIP
Canada: $24.95 US: $24.95 UK: £15.79

Steps to Starting a Successful Import Business

How to Find Products, Bring Them into the Country and Make Money Selling Them

By: Learn2succeed.com Incorporated

Provides the basic knowledge you need to set up and start your own importing business and how to make money selling the products you import. How to find products to import and then do the research to find out whether you can sell them in your local market. How to pay for imports plus freight alternatives for importing your goods and what's involved with customs clearance. Look at some of the advantages of warehousing and bonded warehousing. How to use advertising, direct marketing and e-commerce to sell your imports.

148 pages, softcover, ISBN 978-1-55270-358-8 CIP
Canada: $24.95 US: $24.95 UK: £15.79

Direct Marketing for Beginners

How to Cut Out the Middleman and Sell Direct to Consumers

By: Learn2succeed.com Incorporated

Cut out the middleman and increase your profit margin Review different methods of direct marketing and learn ho to create your own Web site, attract visitors and conduct commerce. Take a look at permission-based e-mail and newsletters. Learn how to sell at auctions on eBay and set u your own eBay store. How to prepare your marketing pla Review the laws and regulations which govern advertising an marketing together with the do-no-call lists.

160 pages, softcover; ISBN 978-1-55270-352-6 CIP
Canada: $24.95 US: $24.95 UK: £15.79

ne-Based Business
Beginners

v to Start a Business on
hoestring from
r Own Home

Learn2succeed.com
›rporated

›u want to run your own business out of your home, this
‹ will provide you with all the information you need to get
ed. Learn about the tax and other advantages of running a
e-based business. But also be alerted to some of the
dvantages including finding good employees and your
liabilities. You will learn where to find products or
ices to sell and how to develop your own products. Then
elp in finding out if there is a market for what you have to
.

pages, softcover; ISBN 978-1-55270-353-3 CIP
ıda: $24.95 US: $24.95 UK: £15.79

Part-Time Business
for Beginners

Successful Ways to
Augment Your Income
While Working for
Someone Else

By: Learn2succeed.com
Incorporated

When you are working for someone else, you probably don't want them to know you have set up your own business, so you will be given some tips on how to keep it secret. How to develop a business idea and figure out if there is a market for the product or service you have selected. Learn how to advertise and sell it. Create your own Web site and open your own Web store. Sell merchandise at eBay auctions or by opening your own eBay store. Learn about legal structures for your part-time business and government sales tax permits.

152 pages, softcover; ISBN 978-1-55270-354-0 CIP
Canada: $24.95 US: $24.95 UK: £15.79

iness Financing
Beginners

ere to Find Money
row Your Business

Learn2succeed.com
›rporated

out how much money you
y need. How to finance
ng different stages of
ness growth; all the way
relatives and friends,to informal investors or angels. What
enture capitalists can offer and hints on negotiating with
including the due diligence process, term sheets and the
agreement. Types of loans and how to prepare before you
y. Take a look at factoring, leasing and how to sell shares
gh an Initial Public Offering (IPO). What's involved in
loyee Share Ownership Plans (ESOPs).

pages, softcover, ISBN 978-1-55270-355-7 CIP
ıda: $24.95 US: $24.95 UK: £15.79

Seven Basic Steps to
Start-Up Business Success

Find a Need, Conduct
Research Prepare Your Plan,
Locate Financing, Start
Operations, Advertise and
Monitor Your Progress

By: Learn2succeed.com Inc.

Find out what people want and where to get great ideas. How to conduct market research and prepare a business plan. Decide between debt or equity. Where to find informal investors or how to get bank financing. Consider the legal form of your business, registration for sales taxes and how to protect your trademarks. Take a look at different marketinga and advertising techniques. How to use permission-based e-mail and e-newsletters. Review your progress and be prepared to change your strategy.

158 pages, softcover, ISBN 978-1-55270-360-1 CIP
Canada: $24.95 US: $24.95 UK: £15.79

For details visit our Canadian Web site: *www.ProductivePublications.ca*
American Web site: *www.ProductivePublications.com*
Order securely online or mail the order form at the end of this catalogue
Phone our Order Desk toll-free at: 1-(877) 879-2669

Streaming Video and Audio for Business

New Ways to Communicate with Your Customers, Employees and Shareholders Over the Internet

By: Learn2succeed.com Incorporated

This timely book looks at new ways for businesses to communicate over the Internet using video and audio. It includes advice on the equipment and software required, together with tips on content creation.

140 pages, softcover; ISBN 978-1-55270-302-1 CIP
Canada: $24.95 US: $24.95 UK: £15.79

Corporate Video Production on a Shoestring

Improve Your Communications with Your Customers, Employees and Shareholders

By: Learn2succeed.com Incorporated

Inexpensive digital camcorders offer great opportunities t improve communications with customers, employees an shareholders. This book covers the equipment and softwa required together with tips on post-production editing and hin on creating great content.

116 pages, softcover; ISBN 978-1-55270-303-8 CIP
Canada: $24.95 US: $24.95 UK: £15.79

e-Business for Beginners

How to Build a Web Site that Brings in the Dough

By: Learn2succeed.com Incorporated

Written for both new and existing businesses, this book introduces you to business on the Internet. It shows you how to create your own Web site, conduct e-commerce, attract customers and get paid. Reviews Web authoring and e-commerce software. How to make your Web site user-friendly and perform search engine optimization.

184 pages, softcover; ISBN 978-1-55270-280-2 CIP
Canada: $29.95 US: $29.95 UK: £18.99

Web Marketing for Small & Home-Based Businesses:

How to Advertise and Sell Your Products Online

By: Learn2succeed.com Incorporated

This book shows you how to advertise and sell your products or services on the Web. Lea the basics of e-commerce and some of the challenges faci online merchants. Find out about search engines and how improve your listings with them. Keep you name in front your customers with permission-based e-mail and electron newsletters. Don't forget the importance of referrals. How use traditional marketing to drive traffic to your site. Find o about the importance of web links and associate programs.

132 pages, softcover; ISBN 978-1-55270-119-5 CIP
Canada: $24.95 US: $24.95 UK: £15.79

For details visit our Canadian Web site: *www.ProductivePublications.ca*
American Web site: *www.ProductivePublications.com*
Order securely online or mail the order form at the end of this catalogue
Phone our Order Desk toll-free at: 1-(877) 879-2669

siness Planning
Beginners

d Out How Much Money
 Will Need to
 Your Business

Learn2succeed.com
orporated

ers your operational, marketing and advertising plans. mines the impact of the Internet. Covers human resource uirements and sub-contracting, automation or puterization to minimize staffing requirements. Set egic objectives and calculate how much money you will d. Covers the manufacturing, service, retail and struction businesses. Printing your plan. How to use it as a ncing document and make effective presentations to ntial investors or lenders.

pages, softcover, ISBN 978-1-55270-356-4 CIP
ada: $24.95 US: $24.95 UK: £15.79

Advertising for
Beginners

Successful Web and
Online Advertising
in the Digital Age

By: Learn2succeed.com
Incorporated

This book emphasises less expensive forms of advertising such as direct mail, print media, yellow pages, signage, trade shows, telemarketing and fax broadcasting. Less emphasis is placed on broadcast media, since this tends to be expensive and often beyond the budget of smaller companies. It shows you how to establish an effective Web presence and how to use offline media to drive traffic to your Web site. Learn how to prepare your advertising plan and the standards and laws which apply to advertising and privacy.

146 pages, softcover; ISBN 978-1-55270-351-9 CIP
Canada: $24.95 US: $24.95 UK: £15.79

ps to Starting a
ession-Proof Business

ere to Find Ideas and
v to Start

Learn2succeed.com
orporated

essions are tough times to start a business and it is rative to choose one that will survive. This book provides of tips on finding areas which will survive and prosper ng a severe economic downturn. It also shows readers how et up their business and keep their start-up costs to a mum.

pages, softcover; ISBN 978-1-55270-381-6 CIP
ada: $24.95 US: $24.95 UK: £15.79

Self-Employment
for Beginners

How to Create Your Own
Job in a Recession

By: Learn2succeed.com
Incorporated

With huge increases in job losses, many people are finding it impossible to find work. Students who are graduating from school, college or university are facing challenges like never before. This book will guide all of them through the process of working for themselves; where to get ideas and how to go about it. It is full of practical tips.

140 pages, softcover; ISBN 978-1-55270-382-3 CIP
Canada: $24.95 US: $24.95 UK: £15.79

For details visit our Canadian Web site: *www.ProductivePublications.ca*
American Web site: *www.ProductivePublications.com*
Order securely online or mail the order form at the end of this catalogue
Phone our Order Desk toll-free at: 1-(877) 879-2669

eBay for Beginners in Canada

How to Buy and Sell at Auctions

By: Learn2succeed.com Incorporated

This timely book will help every Canadian who wants to buy and sell on eBay. It covers issues specific to Canada (unlike most other books which are written for Americans). Selling on eBay is probably one of the easiest ways for you to earn extra income, yet many people do not know how to go about it. This book will provide you with the basic knowledge to get started with a very small investment.

146 pages, softcover, ISBN 978-1-55270-326-7 CIP

Canada: $24.95 US: $24.95 UK: £15.79

Steps to Starting a Successful eBay Business in Canada

Your Path to Financial Independence

By: Learn2succeed.com Incorporated

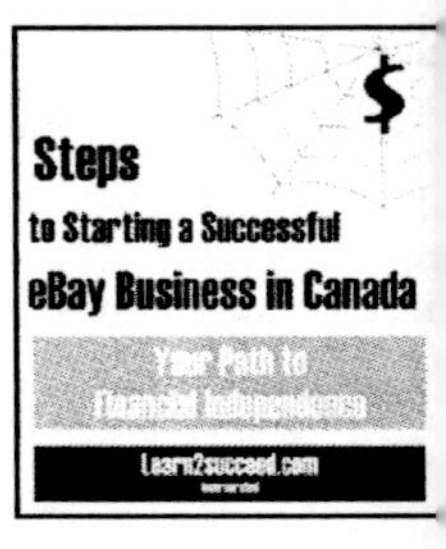

This book will help every Canadian who wants to start business using eBay. It outlines 12 basic steps for success ar covers issues specific to Canada (unlike most other boo which are written for Americans). Welcome to the World Largest Auction! Learn how eBay started; how big it h grown and the basics of selling by auction on eBay.

142 pages, softcover, ISBN 978-1-55270-327-4 CIP

Canada: $24.95 US: $24.95 UK: £15.79

eBay Your Own Home-based Business

Practical Steps to Achieve Financial Independence

By: Learn2succeed.com Incorporated

Written in non-technical language, this book will help you make effective use of eBay to run your own home-based business and make money. You can operate either as a part-time or full-time business. It is written from a Canadian perspective and shows you how to get started with a very small investment. This book starts with an eBay primer and tells you what you can sell and how auctions work. It also shows you the role of eBay in the product cycle

182 pages, softcover, ISBN 978-1-55270-329-8 CIP

Canada: $29.95 US: $29.95 UK: £18.99

Expand Your Canadian Business Using eBay

Everything Managers Need to Know to Start Successfully

By: Learn2succeed.com Incorporated

Written in non-technical language, this book will help every Canadian small busine person make effective use of eBay to increase their sales domestic and foreign markets. It covers issues specific Canadians (unlike most other books which are written f Americans). Selling on eBay is probably one of the easie ways to test new products and sell-off excess inventory end-of-line goods, yet many Canadian businesspeople do n know how to go about it. This book will provide them with t basic knowledge to get started with a very small investment

226 pages, softcover, ISBN 978-1-55270-328-1 CIP

Canada: $34.95 US: $34.95 UK: £21.99

For details visit our Canadian Web site: *www.ProductivePublications.ca*
American Web site: *www.ProductivePublications.com*
Order securely online or mail the order form at the end of this catalogue
Phone our Order Desk toll-free at: 1-(877) 879-2669

Small Business Guide
ncreasing Your Sales
ng eBay

y Ways to Expand into
mestic and Foreign
rkets

Learn2succeed.com
orporated

tten in non-technical language, this book will demonstrate y's role in the product cycle so that every business can e effective use of eBay to increase their sales in domestic foreign markets. How to list products for auction and get . How to open an eBay Store. How to get positive back and integrate eBay into existing operations.

pages, softcover, ISBN 978-1-55270-251-2 CIP
ada: $24.95 US: $24.95 UK: £15.79

Start Your Own
Successful Home-Based
Business Using eBay

Everything You Need to
Know to Get Started

By: Learn2succeed.com
Incorporated

How to start your own full-time or part-time home-based business using eBay. How to register your business, find products, conduct research, list your items for auctions, open an eBay Store and get paid. How to select the right computer hardware and software to help you. The importance of getting positive feedback.

226 pages, softcover, ISBN 978-1-55270-250-5 CIP
Canada: $29.95 US: $29.95 UK: £18.99

xpensive E-Commerce
utions for Small &
ne-Based Businesses:

Don't Have to
nd a Fortune to
rt Selling Online

Learn2succeed.com
rporated

book is a great place to start if you want to learn about pensive e-commerce solutions for your small or e-based business. Find out how to do it inexpensively for t $100 per month. You don't have to spend a fortune to selling online!

pages, softcover;; ISBN 978-1-55270-118--8 CIP
ada: $24.95 US: $24.95 UK: £15.79

Fundamentals of
Effective Online Selling

Use the Power of the
Internet to Increase
Your Sales

By: Learn2succeed.com
Incorporated

How to advertise and sell your products or services on the Web. The basics of online selling and some of the challenges facing online merchants. How to use the Web for your market research and how to prepare your Web marketing plan. Take a tour of inexpensive software to establish your own e-commerce Web site. This book will provide you with all the information you need to start increasing your online sales.

206 pages, softcover; ISBN: 978-1-55270-210-9 CIP
Canada: $29.95 US: $29.95 UK: £18.99

For details visit our Canadian Web site: *www.ProductivePublications.ca*
American Web site: *www.ProductivePublications.com*
Order securely online or mail the order form at the end of this catalogue
Phone our Order Desk toll-free at: 1-(877) 879-2669

Inexpensive Ways to Start Your Own Successful Home-Based E-Commerce Business

How to Set Up Your Business, Select Products & Start Selling Online for Well Under $2,000

By: Learn2succeed.com Inc.

How to start your own online business. Where to get ideas and how to check them out. How to plan your business. How much money will you need? How to get financing. How to advertise and sell your products or services on the Web. The basics of online selling and some of the challenges facing online merchants. How to use the Web for your market research and how to prepare your Web marketing plan.

328 pages, softcover; ISBN 978-1-55270-203-1 CIP
Canada: $39.95; USA: $39.95; UK: £25.19

Steps to Starting a Successful Business in the Digital Age

How to Use the Latest Technology to Turn Your Ideas into Money

By: Learn2succeed.com Incorporated

This book will help everyone who wants to start a business i the digital age. It outlines 10 basic steps for success. How t find ideas for a business, conduct market research, create you own Web site and select your e-commerce software. How t set up your business.

144 pages, softcover, ISBN 978-1-55270-301-4 CIP
Canada: $24.95 US: $24.95 UK: £15.79

eBook Publishing for Beginners

How to Make Money Selling Your Digital Books Online

By: Learn2succeed.com Incorporated

Barclay's Capital has suggested that a quarter of all worldwide book sales in 2015 will be made up of eBooks. This is wake-up call to publishers who are still trapped in the print-on-paper world. It also has ramifications for bookstores, libraries and the book supply chain .

112 pages; Softcover; ISBN: 978-1-55270-456-1 CIP
Canada: $19.99 US: $19.99 UK: $12.59

Stock Market Investing for Beginners

How to Increase Your Wealth in Uncertain Times

By: Learn2succeed.com Incorporated

Written for people who are fed up with the paltry interest the bank pays on their savings accounts as well as those who a sadly disillusioned with the lackluster performance of the investment advisors. Everything you need to know to g started.

164 pages, Softcover; ISBN: 978-1-55270-446-2 CIP
Canada: $24.95 US: $24.95 UK: $15.79

For details visit our Canadian Web site: *www.ProductivePublications.ca*
American Web site: *www.ProductivePublications.com*
Order securely online or mail the order form at the end of this catalogue
Phone our Order Desk toll-free at: 1-(877) 879-2669

ps to Starting and nning a Successful siness in CANADA

Don Lunny

naging your own business can be a rewarding experience survival can be tough in today's economy. This book shows the essential steps to ensure that your business is fitable.

hor, **Don Lunny**, is an experienced business owner and sultant with many years of experience.

pages, softcover ISBN 978-0-920847-85-5 CIP
ada: $34.95

Checklist for Going into Business

By: Don Lunny

Points to create your own checklist to create a profitable business. Starting it is reality. But, there is often a gap between your dream and reality - that can only be filled with careful planning. You need a plan to avoid pitfalls, to achieve your goals and make profits. This guide helps you prepare a comprehensive business plan and determine if your idea is feasible. **Don Lunny** is an experienced business owner and consultant with many years of experience.

53 pages, softcover; ISBN 978-0-920847-86-2 CIP
Canada: $19.95 US: $19.95 UK: £12.59

treet Wise Manager's de to Success he Restaurant siness

Matthew Lallo

could have called this "What they don't teach at The Culinary Institute". As you know, operating a urant is a difficult; even a dangerous business. npetition is fierce and costs keep rising. You are subject to tchwork of government regulations. You will find it a lenge to succeed in this industry, however, this book can you to greatly improve your chances. It provides you with agmatic view of an industry that is unique and it offers you unorthodox (but proven) advice on the subtle art of ival.

pages, softcover; ISBN 978-1-55270-144-7 CIP
ada: $29.95 US: $29.95 UK: £16.98

Food Poisoning and Waterborne Illness

How to Prevent 1.8 Million Deaths Every Year

By: Learn2succeed.com Incorporated

Figures from the World Health Organization and the Government of Ghana, suggest that between 1.8 million and 2.2 million deaths occur every year due to food poisoning and waterborne illness. There are many ways to significantly reduce the death rate. You will find out why governments are reluctant to regulate or take steps to meet the challenge.

116 pages, Softcover; ISBN 978-1-55270-449-3 CIP
Canada: $19.99 US: $19.99 UK: $12.59

Entrepreneurship and Starting a Business

Confederation College Entrepreneurship Series

Entrepreneurship and Starting a Business provides a comprehensive introduction to entrepreneurs and what they do, and is a must-read for anyone who has aspirations to start and run their own business. The book examines entrepreneurs, their values and behaviour, and factors that contribute to their success and failure. It also takes an in-depth look at how they spot business opportunities or come up with business ideas.

110 pages, softcover; ISBN 978-1-55270-090-7 CIP
Canada: $24.95 US: $24.95 UK: £15.79

The Entrepreneur and the Business Idea

Confederation College Entrepreneurship Series

If you ever wondered what entrepreneurs are like; where the look for business ideas and opportunities, and what kinds c thinking and tools some of them use in their approach to possible business start-up, then this introductory book shoul prove very helpful to you. It includes both a self-assessme and a business opportunity assessment tool, and advocates "damage control approach" to getting into business.

50 pages, softcover; ISBN 978-1-55270-089-1 CIP
Canada: $14.95 US: $14.95 UK: £9.49

Business Planning and Finances

Confederation College Entrepreneurship Series

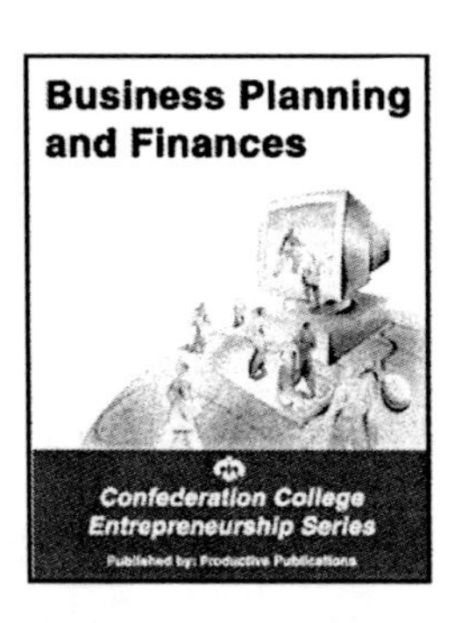

Business Planning and Finances takes a pragmatic and hands-on approach to business planning and financial management, and is written in straightforward language free of technical jargon. It includes a thorough review of the role of planning, the benefits to be realized from planning, and the use of a plan as a management aid.

174 pages, softcover; ISBN 978-1-55270-091-4 CIP
Canada: $34.95 US: $34.95 UK: £21.99

Small Business Finance

Confederation College Entrepreneurship Series

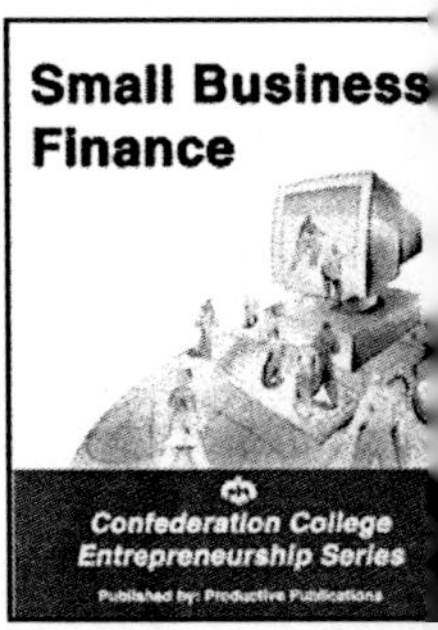

Small Business Finance was designed with the start-up business owner/manager in mi and provides a detailed overview of the organization a operation of a business from a financial perspecti Developed as a combination textbook and workbook, it tal the reader step-by-step through each element of a compan finances from pre-startup costs all the way to record keep and financial monitoring for an established business.

136 pages, softcover; ISBN 978-1-55270-092-1 CIP
Canada: $29.95 US: $29.95 UK: £18.89

For details visit our Canadian Web site: *www.ProductivePublications.ca*
American Web site: *www.ProductivePublications.com*
Order securely online or mail the order form at the end of this catalogue
Phone our Order Desk toll-free at: 1-(877) 879-2669

uth Entrepreneurship

nfederation College
trepreneurship Series

ne of North America's most successful businesses have n started by people between the ages of 15 and 25. If you a young person with a business idea or a desire to start your business then this informative and practical book should "must-read" for you. Learn from the experiences of others improve your prospects for success.

pages, softcover; ISBN 978-1-55270-094-5 CIP
ada: $24.95 US: $24.95 UK: £15.79

Business Relationships – Development and Maintenance

Confederation College
Entrepreneurship Series

The success of any business hinges on the effective management of three critical categories of business relationships. These are a firm's relationships with its customers, with its employees, and with the individuals and organizations that supply it with essential goods and services. This book outlines the nature and role of each type of relationship, and identifies a variety of best practices and operating tools to be employed in the successful development and maintenance of these relationships.

78 pages; softcover; ISBN 978-1-55270-093-8 CIP
Canada: $19.95 US: $19.95 UK: £12.59

coming Successful!

ing Your Home-Based
siness to a New Level

Don Varner

tegies for getting great lts in your home-based ness! How to turn any type business into a CESSFUL business!

- elf-Improvement
- andling Rejections
- lanagement Skills
- 6 Ways to Prospect
- esigning Great Ads
- Self-Motivation
- Hiring Tips
- Motivating Employees
- Closing Sales
- No-Cost Ways to Advertise

pages, softcover; ISBN 978-1-896210-87-2 CIP
ada: $39.95 US: $39.95 UK: £25.19

Start Your Own Business:
Be Your Own Boss!

Your Road Map to
Independence

By: Iain Williamson

Learn from someone who has done it! What it takes! Where to get ideas and how to check them out. How to research the market. Calculate how much money you will really need and where to get it. Growing pains and managing employees... plus lots more. Iain Williamson has run his own businesses for over 35 years and is a consultant. He'll help you with a Road Map to Independence!

208 pages, softcover; ISBN 978-1-896210-96-4 CIP
Canada: $29.95 US: $29.95 UK: £18.99

For details visit our Canadian Web site: *www.ProductivePublications.ca*
American Web site: *www.ProductivePublications.com*
Order securely online or mail the order form at the end of this catalogue
Phone our Order Desk toll-free at: 1-(877) 879-2669

Can You Make Money with Your Idea or Invention?

By: Don Lunny

- Can you exploit it?
- How to produce it.
- Can you make money?
- Where to get help.
- Industrial Design.
- Copyright.
- Points of caution.
- Patent applications.
- Sample licensing agreement.
- Is the idea original?
- How to distribute it.
- Can you protect it?
- A word about patents.
- Trademarks.
- First steps.
- Possible problems.
- What are your chances?

99 pages, softcover; ISBN 978-0-920847-65-7 CIP
Canada: $24.95 US: $24.95 UK: £15.79

The Canadian Business Guide to Patents for Inventions and New Products

By: George Rolston

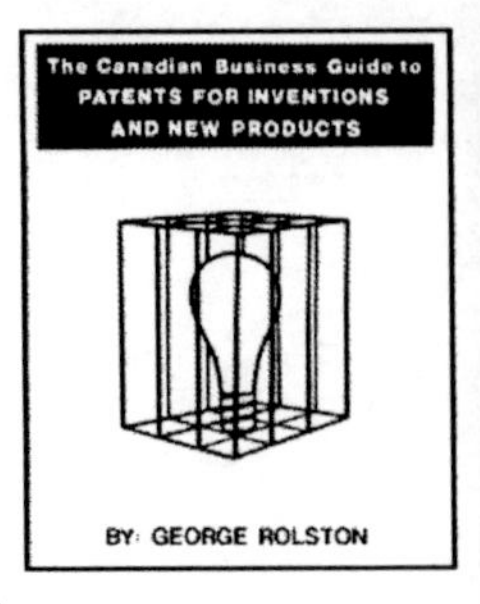

This is your complete reference to patenting around the world. The key elements in the patent process. When to search for earlier patents. When you should file patent applications. The importance of your patent filing date. Understand the critical wording of patent claims. Getting the best out of your patent agent. What the patent office will do for you. What to do if your patent application is rejected. How to go about patenting in foreign countries and how to negotiate a licence agreement. **George Rolston**, is a barrister and solicitor who has specialized in patents for over 30 years.

202 pages, softcover; ISBN 978-0-920847-13-8 CIP
Canada: $48.00

Protect Your Intellectual Property

An International Guide to Patents, Copyrights and Trademarks

By: Hoyt L. Barber & Robert M. Logan

An abundance of information on step-by-step procedures to obtain exclusive protection for unique ideas, inventions, names, identifying marks, or artistic, literary, musical, photographic or cinematographic works. Hoyt Barber is an executive with extensive experience in intellectual property protection. Robert Logan is a practicing U.S. attorney.

305 pages, softcover; ISBN 978-1-896210-95-7 CIP
Canada: $59.95 US: $59.95 UK: £37.79

Evaluating Franchise Opportunities

By: Don Lunny

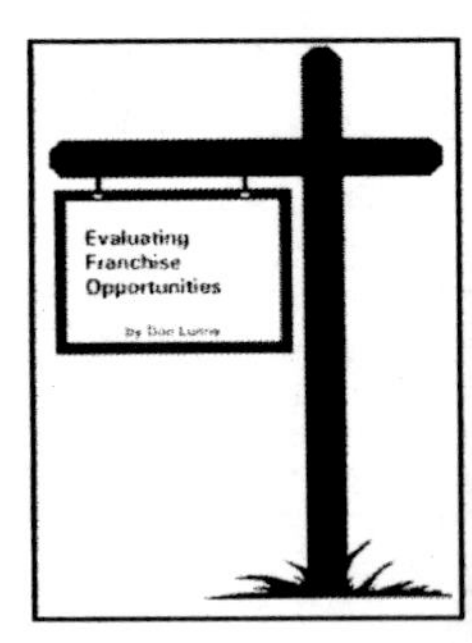

Although the success rate for franchisee-owned businesses is better than for many other start-up businesses, success is not guaranteed. Don't be "pressured" into a franchise that is not right for you. Investigate your options. How to evaluate the business, the franchisor, the franchise package, and yourself. Author and business consultant, **Don Lunny**, shows you how to avoid the pitfalls before you make a franchise investment.

75 pages, softcover; ISBN 978-0-920847-64-0 CIP
Canada: $19.95 US: $19.95 UK: £12.59

For details visit our Canadian Web site: *www.ProductivePublications.ca*
American Web site: *www.ProductivePublications.com*
Order securely online or mail the order form at the end of this catalogue
Phone our Order Desk toll-free at: 1-(877) 879-2669

sic Beancounting

arn to Ape
'rofessional
okkeeper

: T. James Cook, CA

s book is intended to help -accountants understand ic bookkeeping principles procedures so that they can ntain a simple set of accounting records for a small iness. The benefits are up-to-date financial records plus a savings by doing a portion of the work that would erwise be performed by a professional accountant.

pages, softcover; ISBN 978-1-55270-204-8 CIP
ada: $26.95 US: $26.95 UK: £16.99

Gorilla Accounting

How to Survive in a
Jungle of Numbers

By: T. James Cook, CA

Designed to teach the small business owner or manager to read and understand financial statements, and use financial management tools including trend, ratio, and break-even analysis to get maximum information from financial records. Financial and cash flow forecasting are explained and how to use the financial statements effectively. Easy to read, easy to understand, and easy to put into practice.

113 pages, softcover; ISBN 978-1-55270-205-5 CIP
Canada: $26.95

ary Administration

repreneurial Business
isultants of Canada

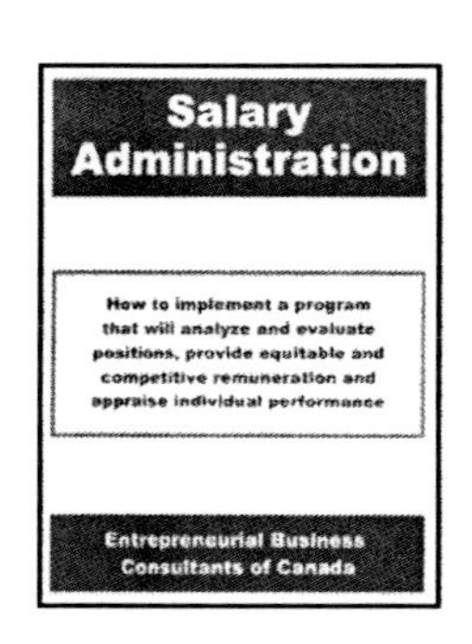

ry Administration Program provides the means for agement to:

Properly analyse and evaluate positions.
Provide equitable and competitive remuneration.
Appraise individual performance in a position.

pages; softcover; ISBN 978-1-55270-085-3 CIP
ada: $39.95 US: $39.95 UK: £25.19

Shoplifting, Security,
Curtailing Crime -
Inside & Out

By: Don Lunny

If you are a shopkeeper or business owner, this practical, hands-on book will alert you to the alarming theft rates you may be exposed to. From petty theft, bad cheques to armed robbery, you get advice on dealing with the situation and how to train staff. Discusses internal theft by employees - how you can recognize it and how to reduce it. If it alerts you to just one problem, it could pay for itself many, many times over.

115 pages; softcover; ISBN 978-0-920847-66-4 CIP
Canada: $29.95 US: $29.95 UK: £18.99

For details visit our Canadian Web site: *www.ProductivePublications.ca*
American Web site: *www.ProductivePublications.com*
Order securely online or mail the order form at the end of this catalogue
Phone our Order Desk toll-free at: 1-(877) 879-2669

Marketing for Beginners

How to Get Your Products into the Hands of Consumers

By: Iain Williamson

Covers the basics of marketing for new entrepreneurs. How to make people aware of your products. How to get them to buy. How to get products into the hands of consumers. Traditional channels of distribution versus direct marketing. One-on-one marketing versus mass marketing. Take a look at the Internet as a marketing tool. Ways to promote and advertise your products. After-sales service and the lifetime value of your customers. Sources of marketing information. The author has been marketing products for 20 years.

215 pages, softcover; ISBN 978-1-896210-97-1 CIP
Canada: $29.95 US: $29.95 UK: £18.99

Marketing Beyond 2000

Why You Will Have to Use the Internet to Market Your Goods or Services in the 21st Century

By: Iain Williamson

The Internet will become an awesome marketing tool in the 21st. Century. Learn how its current limitations are being overcome. Take a look at the future of radio, TV and newspapers.

Glimpse at the marketplace of the future. The author says it's up to you to take advantage of this tremendous marketing tool. Find out how!

194 pages, softcover; ISBN 978-1-896210-66-7 CIP
Canada: $27.95 US: $27.95 UK: £17.69

Successful Direct Mail Marketing in Canada

A Step-by-Step Guide to Selling Your Products or Services Through the Mail

By: Iain Williamson

Techniques to make money in the highly competitive direct mail market. Direct mail as an inexpensive way to reach customers. Ways to keep your costs to a minimum. How to save on postage by using bulk rates. How to get the most out of your computer. The author has over 15 years experience selling by direct mail.

114 pages, softcover; ISBN 978-1-896210-39-1 CIP
Canada: $19.95

Steps to Choosing the Right Computer for Your Home or Business

A No-Nonsense Guide Which Cuts Through All the Hype

By: Learn2succeed.com Incorporated

Read this book **before** you purchase a new computer. It leads you through all factors you should consider. Computer store sales people are more interested in making a commission than on selling you what you need. This book will help. Softcover;

96 pages; Softcover, ISBN: 978-1-55270-454-7 CIP
Canada: $17.99 US: $1799 UK: $11.39

For details visit our Canadian Web site: *www.ProductivePublications.ca*
American Web site: *www.ProductivePublications.com*
Order securely online or mail the order form at the end of this catalogue
Phone our Order Desk toll-free at: 1-(877) 879-2669

e Six Sigma Toolbox

Improvement Tools
d When to Use Them

: Jerry W. Wishes

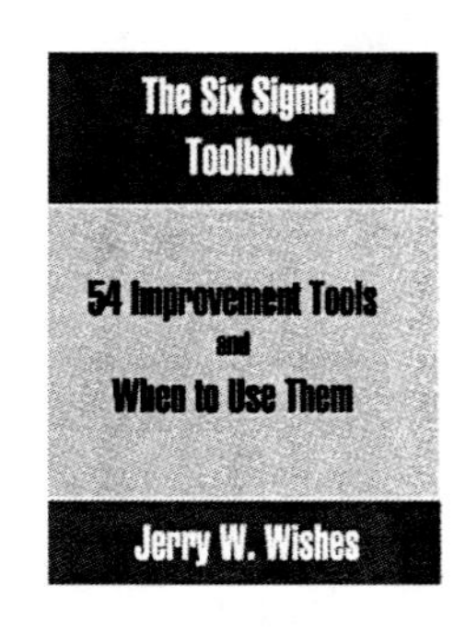

this book Jerry Wishes provides fifty-four improvement ls. They are focused around two concepts. First, Variation vil. Every process has variation but the secret is to restrict o natural causes and then use improvement tools to anage" the variation. Second, use of the tools is not ional. Once you embrace them, you do so forever.

pages, softcover; ISBN 978-1-55270-258-1 CIP
ada: $48.95 US: $48.95 UK: £30.99

Quality in the 21st Century

What You Have to Change to Stay in Business

By: Jerry W. Wishes

Jerry Wishes says: "If you don't 'get it' soon and start doing things differently, you'll be in a down-cycle and never understand why." His hope is that this book will give you some insight and a direction you can use to conquer the challenges ahead.

202 pages, softcover; ISBN 978-1-55270-259-8 CIP
Canada: $39.95 US: $39.95 UK: £25.19

ven Sigma

e Pursuit of Perfection

Jerry W. Wishes

s book makes the case, through the use of a fictional iness story, for the rejection of mediocrity in the corporate ld. The acceptance of 'things just go wrong' is replaced the need to raise the bar on expectations. The story takes e at a high-technology start-up.

pages, softcover; ISBN: 978-1-55270-260-8 CIP
ada: $42.95 US: $42.95 UK: £26.99

Modern Materials Management Techniques:

A Complete Guide to Help You Plan, Direct and Control the Purchase, Production, Storage and Distribution of Goods in Today's Competitive Business Environment –Essentials of Supply Chain Management

By: Paula Mackie **SECOND EDITION**

Covers the entire process of a company's operations relating to the acquisition of goods and services. Written for the public and private sectors as well as college and university educators.

398 pages, softcover; ISBN: 978-1-55270-257-4 CIP
Canada: $74.95 US: $74.95 UK: £47.29

Innovate or Perish!

Seven-Step Innovation Process to Meet the Challenges of Globalization

By: Richard Sussman Sc.D.

A complete manual for manufacturing companies to produce new products and processes that can enhance their competitive position. The process starts with the creation of a strategic innovation plan and then provides a system to evaluate the current products and manufacturing capabilities of a company. Methods to select and execute the new developments in the most effective manner. Outsourcing and executive management are reviewed. Dr. Richard Sussman was one of the top technical leaders in the steel industry.

242 pages, softcover, ISBN 978-1-55270-253-6 CIP
Canada: $49.95 US: $49.95 UK: £21.49

Effective Management:

Interpersonal Skills that Will Help You Earn the Respect and Commitment of Employees

By: Dave Day Ph.D.

Ten key interpersonal skills for the manager... from choosin a leadership style to the day of completing annual performanc evaluations. Contains practical suggestions to increase th productivity and commitment of all employees. Essentia reading for all new managers and a resource for existin managers. Dave Day has over 35 years experience as a manager, consultant and Professor of Management a Columbia College.

180 pages, softcover; ISBN 978-1-896210-99-5 CIP
Canada: $27.95 US: $27.95 UK: £17.69

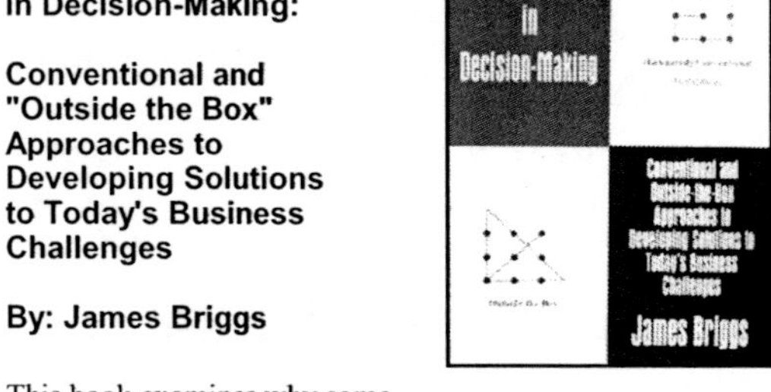

Critical Analysis in Decision-Making:

Conventional and "Outside the Box" Approaches to Developing Solutions to Today's Business Challenges

By: James Briggs

This book examines why some people make good business decisions more effectively, more often, than others. Great leaders in the public service, business, and the non-profit sectors, remind us that an effective decision-making process is the key to solving problems for any organization. Effective organizations search for leaders who have good problem solving skills.

234 pages, softcover; ISBN 978-1-55270-116-4 CIP
Canada: $48.95 US: $48.95 UK: £30.99

Project Management: Welcome Opportunity or Awesome Burden?

By: Robert G. Edwards

This concise, how-to, self-help guide will help both aspirin and practicing project managers. Its content was develope during the author's forty-four years in professional engineerin and project management. The principles and practices that h describes are based on his personal experience and can easil be applied to most simple or complex projects.

170 pages, softcover; ISBN 978-1-55270-086-0 CIP
Canada: $26.95 US: $26.95 UK: £16.99

For details visit our Canadian Web site: *www.ProductivePublications.ca*
American Web site: *www.ProductivePublications.com*
Order securely online or mail the order form at the end of this catalogue
Phone our Order Desk toll-free at: 1-(877) 879-2669

operative Time
nagement

t More Done and
ve More Fun!

: Chance Massaro
Katheryn Allen-Katz

tains the wisdom of the last fifty years of research and ing about time management together with eighteen years king in organizations helping people get the most sfying results. It is intended for people who have goals and t to achieve them. It is interactive and easy to use. The ors are time management experts. Follow the steps which outline in this 224 page workbook and YOUR RESULTS L BE REMARKABLE!

pages, softcover; ISBN 978-1-896210-86-5 CIP
ada: $34.95 US: $34.95 UK: £21.99

THE LEAN OFFICE

How to Use Just-in-Time Techniques to Streamline Your Office

By: Jim Thompson

This book is for everyone who works in an office. Find out how to foster and nurture employee involvement and put excitement back into continuous improvement. Get the tools needed to improve office productivity. Most importantly, reduce employee stress and frustration, while improving productivity. Find out how this happens with employees, not to employees! Jim Thompson is a lean production consultant who studied these systems first-hand while with GM and Toyota in California.

138 pages, softcover; ISBN 978-1-896210-41-4 CIP
Canada: $24.95 US: $24.95 UK: £15.79

N PRODUCTION

v to Use the Highly
ective Japanese
cept of Kaizen to
rove Your Efficiency

Jim Thompson

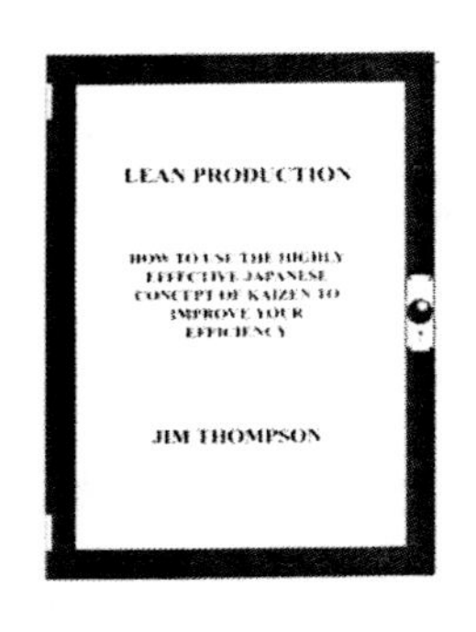

n specific techniques and behaviours to improve your tiveness. Find out about a system that has been used very tively at the organizational level for over forty years. or, **Jim Thompson** has held senior management positions General Motors and the Walker Manufacturing Company.

pages, softcover; ISBN 978-1-896210-42-1 CIP
da: $24.95 US: $24.95 UK: £15.79

LEAN PRODUCTION FOR THE OFFICE

Common Sense Ideas To Help Your Office Continuously Improve

By: Jim Thompson

More ideas for everyone who works in an office:

- Be idea-driven
- Reduce frustration
- Add value
- Let others benchmark you

How to use employees' creativity and ingenuity. Employees' feelings **do** count! Author, **Jim Thompson**, is the guru of applying lean production to the office environment.

136 pages, softcover; ISBN 978-1-55270-025-9 CIP
Canada: $24.95 US: $24.95 UK: £15.79

For details visit our Canadian Web site: *www.ProductivePublications.ca*
American Web site: *www.ProductivePublications.com*
Order securely online or mail the order form at the end of this catalogue
Phone our Order Desk toll-free at: 1-(877) 879-2669

Speak Up!

Helpful Tips for Business People who Need to Speak in Public

By: T. James Cook, CA

Helps business people improve their speaking skills and become good oral communicators. Includes an assessment of present skills together with practice sessions. Better speaking skills will lead you to more responsibility, authority, and material benefits. Confidence in your speaking skills will result in reduced stress levels when you know that you are going to have to speak, or when you are in a potential speaking situation. Ideas that are well articulated are always given more weight and make you more successful.

110 pages, softcover; ISBN 978-1-55270-256-7 CIP
Canada: $24.95 US: $24.95 UK: £15.79

Quick Fixes for Business Writing

An Eight-Step Editing Process to Find and Correct Common Readability Problems

By: Jim Taylor

Breaks down the editorial process into a series of tasks which are designed to improve the readability of the final product. It will be invaluable to you; regardless of whether you are a novice or a proficien editor. Author, Jim Taylor, has taught Eight-Step Editing fo 18 years and clients for his workshops include the Editor' Association of Canada and the Ontario Cabinet Office.

156 pages, softcover; ISBN 978-1-55270-252-9 CIP
Canada: $24.95 US: $ 24. 95 UK: £15.79

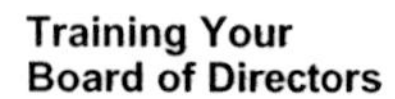

Training Your Board of Directors

A Manual for the CEOs, Board Members, Administrators and Executives of Corporations, Associations, Non-Profit and Religious Organizations

By: ArLyne Diamond, Ph.D.

For more than ten years now the author has been training boards of directors of organizations of all kinds, from religious organizations to fast-growing high tech companies. This manual is different from most board training books. It is a combination of short informative pieces and a series of interactive exercises designed to enable the participants to actively reach the desired conclusions rather than being lectured to, or corrected by "the expert."

350 pages, softcover; ISBN 978-1-55270-207-9 CIP
Canada: $39.95 US: $39.95 UK: £25.19

Leadership with Panache

52 Ways to Set Yourself Apart as a Dynamic Manager

By: Jeff Jernigan

This book cuts to the underbelly of leadership in the modern organization. Divided into "Ways" so that you can select one topic each week of the y for group discussion with your management and supervis associates. Poses hard hitting questions for considerati Author, **Jeff Jernigan**, has over 25-years experience as organizational development specialist providing compan support in creating, continuing and capitalizing on change. is the recipient of numerous industry awards.

180 pages, softcover; ISBN 978-1-55270-081-5 CIP
Canada: $29.95 US: $29.95 UK: £18.99

w to Deliver Excellent
stomer Service

tep-by-Step Guide
Every Business

: Julie Olley

re-designed workbook roach for businesses that h to develop, implement, analyse and follow-up customer ice projects. Step-by-step "HOW TO:" ideas and sample nats are included. The suggestions can be implemented r time. Author, **Julie Olley**, was formerly National nager of Quality Assurance with a major international el organization. She has designed several curricula for The adian School of Management and International Business.

pages, softcover; ISBN 978-1-55270-045-7 CIP
ada: $26.95 US: $26.95 UK: £16.99

Time Management for
Beginners

How to Get the Most Out
of Every Day

By: Learn2succeed.com
Incorporated

There is a saying that some people count time, while others make time count. This book is about making time count. It's about managing your time effectively so that you can get the most out of each and every day of your life.

114 pages; Softcover, ISBN: 978-1-55270-453-0 CIP
Canada: $19.99 US: $19.99 UK: $12.59

ital Photography
Beginners

v to Create Great
otos for Fun or Profit

Learn2succeed.com
orporated

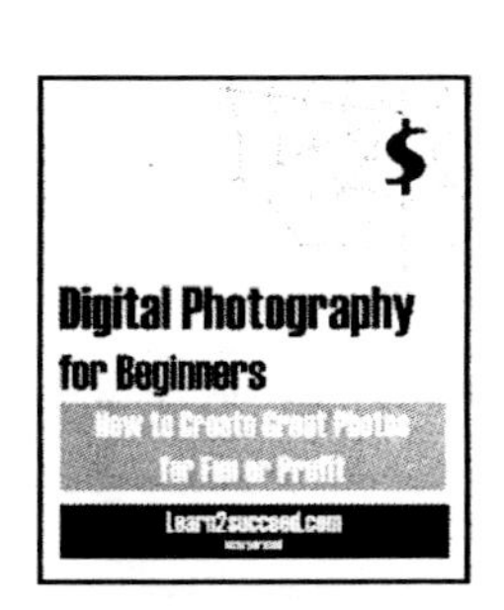

u want to become a digital shutterbug, the place to start is eading this excellent introduction which tries to explain ything in plain, non-technical terms.

ages; Softcover; ISBN: 978-1-55270-447-9 CIP
ada $17.99 US: $1799 UK: $11.39

Timeless Strategies to
Become a Successful
Entrepreneur

By:
Lawrence Scott Troemel

This book is all about starting, building, and managing a small business. The approaches covered in this book have been successfully implemented for decades and will continue to be viable well into the future. Every entrepreneur will benefit from the advice in this very readable book. It is also full of interesting anecdotes.

208 pages; softcover; ISBN 978-1-55270-046-4 CIP
Canada: $29.95 US: $29.95 UK: £18.99

For details visit our Canadian Web site: *www.ProductivePublications.ca*
American Web site: *www.ProductivePublications.com*
Order securely online or mail the order form at the end of this catalogue
Phone our Order Desk toll-free at: 1-(877) 879-2669

Desktop Publishing for Beginners

How to Create Great Looking Brochures, Books and Documents

By: Learn2succeed.com Incorporated

This book will introduce you to desktop publishing and shows you how you can create your own brochures, books and documents. It is s, although some are available for the Macintosh.

114 pages; Softcover; ISBN: 978-1-55270-455-4 CIP
Canada: $19.99 US: $19.99 UK: $12.59

How to Buy or Sell a Business

Questions You Should Ask and How to Get the Best Price

By: Don Lunny

The decision to buy or sell a business requires carefu consideration. It may affect the course of the participant future lives. Yet a surprising number of owners rush int transactions without adequate preparation. Find out how to se the price, locate prospects, evaluate offers, close deals an finance purchases. Author, **Donald Lunny**, has many years c business experience and has been involved with the purchas and sale of many businesses.

134 pages, softcover; ISBN 978-1-896210-98-8 CIP
Canada: $24.95 US: $24.95 UK: £15.79

Tips for Entrepreneurs

How to Meet the Challenges of Starting And Managing Your Own Business

By: Henry Kyambalesa

This book is the culmination of a 3-year research study into the challenges faced by entrepreneurs when they become their own boss. Tips for those about to start a business & tips for those already in business. Decide whether self-employment is for you. Practical advice on getting started. The skills you will need. Henry Kyambalesa is a tenured lecturer in Business Administration. He holds B.B.A., M.A., and M.B.A. degrees.

194 pages, softcover ISBN 978-1-896210-85-8 CIP
Canada: $26.95 US: $26.95 UK: £16.99

Work from Your Home Office as an Independent Contractor

A Complete Guide to Getting Started

By: Chantelle Sauer

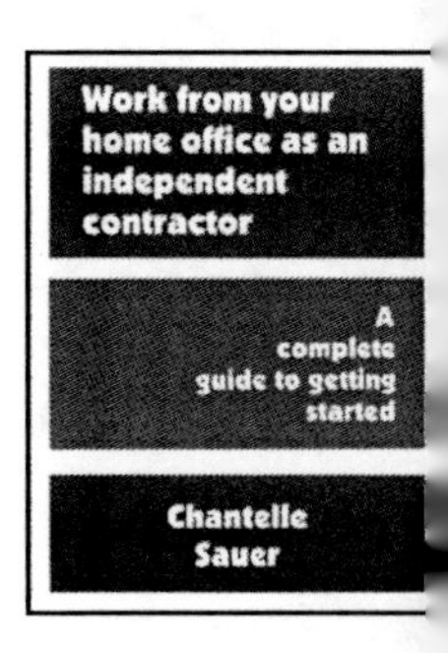

An independent contractor is someone who works from his her home or home office e.g., consultants, entrepreneur business owners, freelancers and outsourcers. Learn about th advantages and disadvantages as well as the legal obligation Also get many ideas on how to become an independe contractor. Author, **Chantelle Sauer**, has spent four years an independent contractor. She knows from first-har experience how to get work.

166 pages, softcover; ISBN 978-1-55270-077-8 CIP
Canada: $24.95 US: $24.95 UK: £15.79

For details visit our Canadian Web site: *www.ProductivePublications.ca*
American Web site: *www.ProductivePublications.com*
Order securely online or mail the order form at the end of this catalogue
Phone our Order Desk toll-free at: 1-(877) 879-2669

e Basics for
les Success

Essential Guide for New
les Representatives,
trepreneurs and
siness People

: Bill Sobye

introductory book which covers the basic points on how to:

- Find customers
- Study your prospects
- Dress for success
- Handle "the butterflies"
- Set goals
- How to include humour
- Success and rejection
- Business versus pleasure

Sobye has 28 years of experience as a Sales Manager.

pages; softcover; ISBN 978-1-896210-65-0 CIP
ada: $24.95 US: $24.95 UK: £15.79

Bulletproof Salesman

A Lively Guide to Enhance Your Sales Techniques

By: Steven Travis Smith & Bruce D. Seymour

Strategies to Help You Bridge the Gap Between Textbook Training and the Real World

A humorous, yet practical guide written using a tag-team approach between the authors. It explains how they've completely screwed-up over the years avoided making the same mistakes again. Learn from their failures as well as their victories. They explain how to reach absolutely anyone, evade the traps constructed to keep salespeople out, and how to instantly detect deception during negotiations.

231 pages, softcover; ISBN 978-1-55270-209-3 CIP
Canada: $29.95 $US: $29.95 £18.99

tware for Small
siness 2011 Edition

ndows & Vista
grams to Help you
rove Business
ciency and
ductivity

Iain Williamson

iews of 240 programs for
and experienced users. Covers operating systems, word
essing, desktop publishing, voice dictation, graphics,
al photography, digital video & audio, spreadsheets,
unting, databases, contact management, communications,
net software, security and virus protection.

pages, softcover; ISBN978-1-55270-405-9 CIP
N: 1492-384X Canada: $79.95 US: $79.95 UK: £50.49

Learn UNIX in Fifteen Days

By: Dwight Baer and Paul Davidson

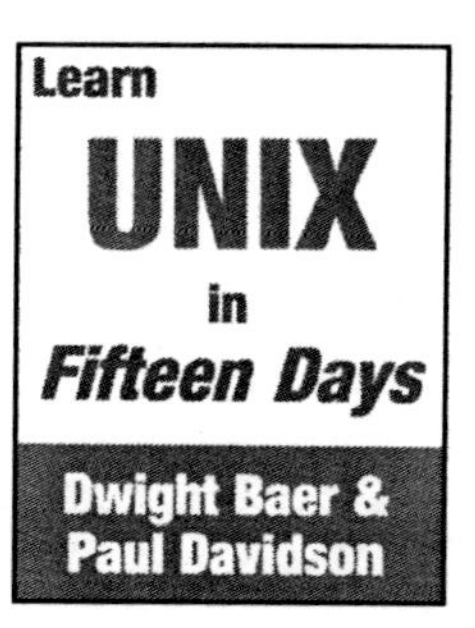

This book was written out of the need for a text which presents the material which is actually taught in a typical UNIX course at the college level. It is not intended to replace a comprehensive UNIX manual, but for most students who have not yet spent five years learning all the "eccentricities" of the UNIX Operating System, it will present all they need to know (and more!) in order to use and support a UNIX system.

176 pages, softcover; ISBN: 978-1-55270-087-7 CIP
Canada: $34.95 US: $34.95 UK: £21.99

For details visit our Canadian Web site: *www.ProductivePublications.ca*
American Web site: *www.ProductivePublications.com*
Order securely online or mail the order form at the end of this catalogue
Phone our Order Desk toll-free at: 1-(877) 879-2669

Management During an Economic Crisis

Best Practices for Small Business Survival in a Recession

By: Robert Papes

Best practices which are vital to every small business to help them survive the current recession because "hope" of better times is not a viable strategy. This book is all meat and potatoes with no filler. Author, Robert Papes, is a consultant with many years of experience in helping businesses in difficulty.

182 pages, softcover; ISBN 1-55270-384-7 CIP
Canada: $29.95 US: $29.95 UK: £18.99

MAKE IT! MARKET IT! BANK IT!

Over 100 Ways to Start Your Own Home-Based Business

By: Barbara J. Albrecht

This book is about starting your own home-based business. It's also about earning extra money when your wages don't stretch far enough. Money fc vacations and education often fall through the cracks in you financial plans and you may find that you need a secon income. Newspaper columnist, Barb Albrecht, has assemble these 100 great ideas to help you put cash into your "mone jar". If you're looking to run your own part-time business c start a new career as owner of your own enterprise....you mus read this book.

144 pages, softcover; ISBN 978-1-55270-145-4 CIP
Canada: $24.95 US: $24.95 UK: £15.79

Your Homebased Business Plan

-Also-

Working With Your Banker

By: Donald Lunny

SECTION I - The Business Plan for Homebased Business: a step-by-step guide to writing it.

SECTION II - Working with your Banker: the fundamentals of borrowing and how they affect you.

Donald Lunny: an entrepreneur and consultant with many years experience in organizing and restructuring companies.

52 pages, softcover; ISBN 978-0-920847-35-0 CIP
Canada: $14.95 US: $14.95 £9.49

THE NET EFFECT

Will the Internet be a Panacea or Curse for Business and Society in the Next Ten Years?

By: Iain Williamson

Are you ready for the greatest change to business & socie since the Industrial Revolution? Examine the world ten yea from now when entire sectors of the economy may eliminated and others will be born. Find out who will be t winners and losers and how it will affect you. Prepare for t dramatic changes that are coming!

244 pages, softcover, ISBN 1-896210-38-4; CIP
Canada: $29.95 USA: $21.95 UK: £18.99

For details visit our Canadian Web site: *www.ProductivePublications.ca*
American Web site: *www.ProductivePublications.com*
Order securely online or mail the order form at the end of this catalogue
Phone our Order Desk toll-free at: 1-(877) 879-2669

ath by Food

ıy More People in North
ıerica Die By Food
isoning than Were
rdered in 9/11

: Iain Williamson

ou've ever had food poisoning, you're certainly not alone. nillion Americans and 12 million Canadians suffer from a dborne illness **every year**. Food poisoning is now one of leading causes of illness in both Canada and the United es. Deaths due to foodborne pathogens total about 5,000 **ually** for the US. This no-nonsense book claims that the d you eat may be neither safe nor healthy and suggests t you can do about it.

pages, softcover; ISBN 978-1-55270-383-0 CIP
ada: $24.95 US: $24.95 UK: £15.79

Yoga for Mind, Body and Spirit

Details of Practices That Will Help Your Health, Psychological and Spiritual Well-Being

By: Dr. John R.M. Goyeche

Detailed descriptions of certain yoga practices for physical, mental and spiritual development; for different-aged people; for different situations and for people with certain health concerns. Dr. Goyeche has studied and taught yoga on three continents and in many settings for over 30 years.

188 pages, softcover; ISBN 978-1-55270-056-3 CIP
Canada: $24.95 US: $24.95 UK: £15.79

ıning Over Depression

-Energetic Therapy
)vercome Sadness,
ır and Anger

Dr. John R.M. Goyeche

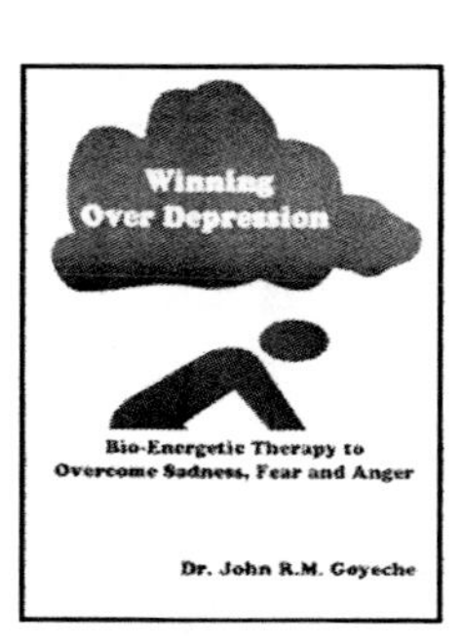

aightforward book that discusses the causes of depression. rs practical answers to overcome it. Written for the general ic and also practitioners of psychiatry, medicine and h care. Dr. John R.M. Goyeche has 25 years of clinical rience in hospitals, mental health centres and bilitation clinics..

pages, softcover; ISBN 978-1-55270-051-8 CIP
ıda: $24.95 US: $24.98 UK:£15.79

Benefit From Hypnosis, Hypnosis by Telephone and Self-Hypnosis

How to Improve Your Self-Esteem, Creativity and Performance as well as Your Spiritual, Physical and Mental Well-Being

By: Dr. John R.M. Goyeche

Hypnosis can help you with:

Habits & Addictions	Fears & Anxieties
Self-Esteem & Depression	General Medical Problems
Creative Activity	Spirituality
Memory Retrieval	Performance Enhancement

Dr. Goyeche is a member of the Canadian Society of Clinical Hypnosis, a Fellow of the International College of Psychosomatic Medicine, a member of the International Institute for Bio-Energetic Analysis.

212 pages, softcover; ISBN: 978-1-55270-050-1 CIP
Canada: $21.95 US: $21.95 UK: £13.89

For details visit our Canadian Web site: *www.ProductivePublications.ca*
American Web site: *www.ProductivePublications.com*
Order securely online or mail the order form at the end of this catalogue
Phone our Order Desk toll-free at: 1-(877) 879-2669

Slot Machines: Fun Machines or Tax Machines?

A Technician Reveals the Truth About One-Armed Bandits

By: Ian B. Williams

How slot machines work and how to play them. Covers the pay-out systems. Helps you have a better casino experience. Also examines the social implications of slot machines in our society; both the positive and negative. **Ian B. Williams** is a certified electronics technician and a trained slot technician, who worked for several years in the casino industry.

134 pages, softcover; ISBN 978-1-55270-049-5 CIP
Canada: $24.95 US: $ 24.95 UK: £15.79

Make Money Trading Options

How to Start Immediately

By: Jason Diptee

Want to invest in an expensive stock, the Japanese Yen or the DOW but only have $200- $300 to invest? Option tradin allows you to enter these markets to take advantage o investment opportunities that would otherwise requir thousands of dollars. This book will teach beginners how t participate in the largely untapped and unknown area o investing that can generate profits in a matter of weeks. Jaso Diptee holds an MBA and is an experienced seminar leader o the subject of option trading.

116 pages, softcover; ISBN 978-1-55270-148-5 CIP
Canada: $24.95 US: $24.95 UK: £15.79

Dollars to Donuts

A Personal Wealth Management Model for Canadians

By:Daniel Kesselring

Five Easy Steps that Can Change Your Financial Direction Today

This book provides a unique unbiased perspective from outside the financial services and wealth management industries. It is a product of research, personal observation and a lifetime of trial-and-error experience that led to a system of money management that has served the author extraordinarily well over the years.

157 pages, softcover; ISBN 978-1-55270-208-6 CIP
Canada: $24.95

Gold Investing for Beginners

An Opportunity for Huge Gains or a Bubble About to Burst?

By: Learn2succeed.com Incorporated

Find out about the major factors which influence the price gold and determine for yourself whether a gold bubble h developed and is about to burst or whether investing in go still offers an opportunity to make huge gains.

134 page; Softcover, ISBN: 978-1-55270-445 CIP
Canada $19.99 US: $19.99 UK: $12.59

For details visit our Canadian Web site: *www.ProductivePublications.ca*
American Web site: *www.ProductivePublications.com*
Order securely online or mail the order form at the end of this catalogue
Phone our Order Desk toll-free at: 1-(877) 879-2669

, You Wanna a Millionaire...

James P. Johnson

; book provides you with ep-by-step guide to developing a personalized financial a that will help you build wealth. The techniques are very ple to understand and the author has done a great job in aining the basic concepts in a straightforward way. He has aded many tables that you can immediately use in creating r own wealth-building plan.

pages, softcover; ISBN: 978-1-55270-088-4 CIP;
ada: $36.95 US: $36.95 UK: £23.29

Short Cut to Easy Street

How to Get Money in Your Mailbox Every Day, Plus Automatic Income for the Rest of Your Life

By: Stephen W. Kenyon

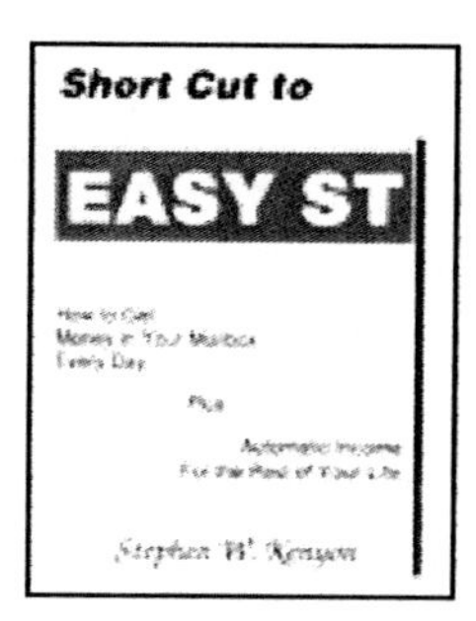

A great book on self-motivation, direct mail, self-publishing, marketing/advertising/promoting and network marketing. Study and learn the details of Stephen Kenyon's fascinating system for attracting wealth and success. He shares with you the inside trade secrets and techniques which he learned over a 30-year period.

244 pages, softcover; ISBN 978-1-55270-057-0 CIP
Canada: $37.95 US: $37.95 UK: £23.99

v to Write a Million Dollar enture Novel

'el Writing as a fitable Profession

Dr. Ray Mesluk

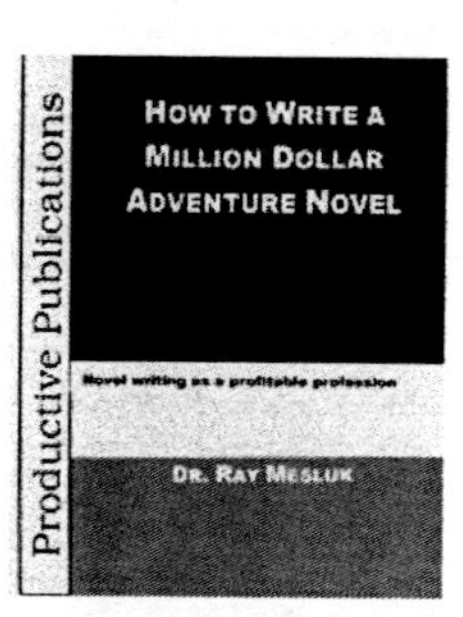

uctured approach to writing your novel quickly and easily. ter the techniques of novel writing and turn them into a table career.

pages, softcover; ISBN 978-1-55270-001-3 CIP
da: $34.95 US: $34.95 UK: £21.99

The "Please" & "Thank You" of Fundraising for Non-Profits:

Fifteen Essential Ingredients for Success

By: ArLyne Diamond, Ph.D.

The "Please" and "Thank You" of Fundraising for Non-Profits
Fifteen Essential Ingredients for Success
ArLyne Diamond, Ph.D.

This book will show you how to successfully raise funds for non-profits; whether you are a member of a national organization, or a small community association. The author, Dr. ArLyne Diamond, has many years of experience with non-profits. She says raising money is an art form rather than a science. Her book shows you how to do it.

126 pages, softcover; ISBN 978-1-55270-261-1 CIP
Canada $24.95 $US: $24.95 UK:£15.77

For details visit our Canadian Web site: *www.ProductivePublications.ca*
American Web site: *www.ProductivePublications.com*
Order securely online or mail the order form at the end of this catalogue
Phone our Order Desk toll-free at: 1-(877) 879-2669

ORDER FORM

Qty.	Title	Price
	ADD Postage: $12.95 first title within Canada or $14.95 to USA	
	ADD $2.25 Postage per title thereafter in Canada or $3.75 to USA	
	SUB-TOTAL	
	ADD 5% HST - Canadian Residents Only (others EXEMPT)	
	TOTAL	

Name____

Organization____

Street____

City/Town___________________________State/Prov_____ Zip/Postal Code__________

Phone___________________Fax___________________

☐ Cheque ☐ VISA ☐ MasterCard ☐ American Express

VISA MasterCard AMEX

Credit Card Orders: can be faxed to: + (416) 322-7434

Card Number____

Expiry Date(Month/Year)____

Cardholder Signature____

Mail to: **Productive Publications**
PO Box 7200, Stn. A, Toronto, ON M5W 1X8 Canada
Order Desk toll-free: 1-(877) 879-2669 Fax: (416) 322-7434
Order Securely Online at: www.ProductivePublications.ca